EAI/Springer Innovations in Communication and Computing

Series Editor
Imrich Chlamtac, European Alliance for Innovation
Ghent, Belgium

The impact of information technologies is creating a new world yet not fully understood. The extent and speed of economic, life style and social changes already perceived in everyday life is hard to estimate without understanding the technological driving forces behind it. This series presents contributed volumes featuring the latest research and development in the various information engineering technologies that play a key role in this process. The range of topics, focusing primarily on communications and computing engineering include, but are not limited to, wireless networks; mobile communication; design and learning; gaming; interaction; e-health and pervasive healthcare; energy management; smart grids; internet of things; cognitive radio networks; computation; cloud computing; ubiquitous connectivity, and in mode general smart living, smart cities, Internet of Things and more. The series publishes a combination of expanded papers selected from hosted and sponsored European Alliance for Innovation (EAI) conferences that present cutting edge, global research as well as provide new perspectives on traditional related engineering fields. This content, complemented with open calls for contribution of book titles and individual chapters, together maintain Springer's and EAI's high standards of academic excellence. The audience for the books consists of researchers, industry professionals, advanced level students as well as practitioners in related fields of activity include information and communication specialists, security experts, economists, urban planners, doctors, and in general representatives in all those walks of life affected ad contributing to the information revolution.

Indexing: This series is indexed in Scopus, Ei Compendex, and zbMATH.

About EAI - EAI is a grassroots member organization initiated through cooperation between businesses, public, private and government organizations to address the global challenges of Europe's future competitiveness and link the European Research community with its counterparts around the globe. EAI reaches out to hundreds of thousands of individual subscribers on all continents and collaborates with an institutional member base including Fortune 500 companies, government organizations, and educational institutions, provide a free research and innovation platform. Through its open free membership model EAI promotes a new research and innovation culture based on collaboration, connectivity and recognition of excellence by community.

Andreia de Bem Machado
Maria José Sousa • Francesca Dal Mas
Silvana Secinaro • Davide Calandra
Editors

Digital Transformation in Higher Education Institutions

Springer

Editors
Andreia de Bem Machado
Federal University of Santa Catarina
Florianópolis, Brazil

Francesca Dal Mas
Ca' Foscari University of Venice
Department of Management
Venice, Italy

Davide Calandra
Department of Management
"Valter Cantino"
University of Turin
Torino, Italy

Maria José Sousa
Instituto Universitário de Lisboa
Lisboa, Portugal

Silvana Secinaro
Department of Management
University of Turin
Torino, Italy

ISSN 2522-8595 ISSN 2522-8609 (electronic)
EAI/Springer Innovations in Communication and Computing
ISBN 978-3-031-52295-6 ISBN 978-3-031-52296-3 (eBook)
https://doi.org/10.1007/978-3-031-52296-3

This Springer imprint is published by the registered company Springer Nature Switzerland AG
The registered company address is: Gewerbestrasse 11, 6330 Cham, Switzerland

Paper in this product is recyclable.

Prologue

"The Times They are a-Changing..."
—Bob Dylan

The title of this prologue is a song written by Bob Dylan and was released as the title track of his 1964 album of the same name. It became famous during a time of countercultures, political changes including the civil rights movement and the beginning of environmental consciousness.

During that time the youth were proclaiming a world of "peace, love and hope". They were hoping for a world that would stop using them as machines in factory production lines and stop using them as soldiers in wars. They hoped for a world where they could live in peace and harmony with each other and with nature.

However, their plight was not heard at that time. The education models continued as before. Students were expected to sit in rows and expected to watch demonstrations and to memorize and regurgitate the knowledge they learned. Continuously being kept in a system that did not promote creative and out-of-the-box thinking.

Fortunately, world and religious leaders became aware of the need for change and organizations such as UNESCO (1945), the World Bank (1944), Let No Child Left Behind (NCLB) (1965), Bread for the World (1974) as well as the European Union (1992) and the African Union (2002) were founded. Political systems and economical systems transitioned from socialist to liberal-democratic systems (Fukuyama, 2020), promoting education to the poor and uneducated.

However, even though education became available to more children, the actual education systems stayed as traditional as before. Churning out young people for the work force to fill jobs applying the same behaviorist pattern they had learned.

And then 20 years later something happened in the field of communication information technology (ICT). The World Wide Web or Internet as we know it was invented by a British computer scientist, Tim Berners-Lee, using a global hypertext document system that could establish communication which aimed at uniting researchers and would offer open access to all (Berners-Lee, 1989).

From then on more Education systems started incorporating computer technology as they could see the value of students being part of their own learning and searching for information on their own (Agarwal & Day, 1998; Lenhart et al., 2001; Ramani, 2015).

Another triumph for the hope of change came at the turn of the century when technology became mobile and teaching and learning became ubiquitous and self-directed (Kinshuk & Graf, 2012). Teachers started flipping their classrooms (Bishop & Vergelegen, 2013); changing them to group seating and moving (Renard, 2020; Alik & Arabi, 2021) from teacher-centered learning to student-centered learning (Garrett, 2008). However, many educators and policy makers still had their reservations to move from their traditional approach (Peters, 2010).

At the same time and specifically since November 2022 the world was introduced to Internet-based intelligence applications that would change education, economies and policies into a new era, namely, Generative Artificial Intelligence (Agrawal et al., 2019; Williams, 2023; Chan, 2023).

Besides existing search engines such as Google and Bing, these AI systems offer opportunities to lecturers, students, businesses, architects, artist, doctors and even politicians to research and assist them with every day, tedious tasks and turn them out within minutes. Tasks such as designs, documentation, contracts, diagnosis curricula, writing letters, creating images and many more.

Do the changes mean that human kind is getting closer to achieving a society where humans can learn more about each other by having access to more knowledge, resources and communication opportunities? Will these technological advances introduce a new era where social and emotional as well as economic wellbeing can be addressed and solved?

The title of this book *Digital Transformation in the Higher Education Institution*, as presented in each chapter, indicates the impact of transitioning and integrating technology into Higher Education. The study also indicates how these changes in the Higher Education Landscape are of utmost importance for student's preparedness to transition into the world of work. These changes offer knowledge and opportunities in education and thereby opportunities to improve prosperity for all.

For example, using technology as simple as a smartphone can seamlessly connect us from the sofa of our home or even from a taxi to AI. It can also connect us with every service we need or can contribute to (Hambrock & De Villiers, 2023). Thus, transforming learning and teaching as well as our lifestyle to a seamless experience.

Other possible advantages of using generative AI could be to cut down on unnecessary time spend on tedious work so that school and work could start an hour later and end an hour earlier. Families could spend more time together and more time could be available for social and community gatherings to improve general wellbeing of humans.

Would this mean that a future where peace, love and hope could be achieved because humanity no longer has to fight over resources on a micro-level as they can rather work together for a world where the global environment would thrive and be alive for future generations? Could this change restore our natural environment so that it could in turn take care of humanity again?

Perhaps soon, these changes could be added as a last verse to Bob Dylan song: "The times have changed from war, hunger and poverty. To peace, love, hope and prosperity for all".

Dr. Helga Hambrock
Chair of International Research Network
IRN-ISSE.org
Chicago, USA

References

Agarwal, R., & Day, A. E. (1998). The impact of the Internet on economic education. *The Journal of Economic Education, 29*(2), 99–110.

Agrawal, A., Gans, J., & Goldfarb, A. (2019). Economic policy for artificial intelligence. *Innovation policy and the economy, 19*(1), 139–159.

Alik, R., & Arabi, M. (2021). *Investigating the impact of classroom seating arrangement on EFL learners' motivation*. Doctoral dissertation, Université Ibn Khaldoun-Tiaret.

Anonymous. (2023a). *The impact of the internet on education: How the web changes the classroom*. Retrieved from https://mspoweruser.com/the-impact-of-the-internet-on-education-how-the-webchanges-theclassroom/#:~:text=With%20a%20few%20clicks%2C%20students%20can%20find%20answers%20to%20almost,process%20more%20engaging%20and%20effective

Anonymous. (2023b). *World Bank – What we do*. Retrieved from https://www.worldbank.org/en/about#:~:text=End%20extreme%20poverty%20within%20a%20generation%20and%20boost%20shared%20prosperity.&text=Mission,The%20World%20Bank%20Group%20has%20two%20ambitious%20goals%3A,generation%20and%20boost%20shared%20prosperity

Anonymous. (2023c). *About the African Union*. Retrieved from https://au.int/en/overview#:~:text=The%20African%20Union%20(AU)%20was,to%20build%20on%20its%20work

Anonymous. (2023d). *Bread for the world*. Retrieved from https://www.bread.org/history/#:~:text=In%20the%201970s%2C%20Rev.,and%20poverty%20in%20his%20neighborhood

Anonymous. (2023e). *UNESCO's history*. Retrieved from https://www.unesco.org/en/brief#:~:text=UNESCO's%20History&text=New%20governments%2C%20including%20that%20of,1%20to%2016%20November%201945

Berners-Lee, T. (1989). Tim Berners-Lee, Bishop, J., & Verleger, M. A. (2013, June). The flipped classroom: A survey of the research. In 2013 ASEE annual conference & exposition (pp. 23-1200)

Chan, C. K. Y. (2023). A comprehensive AI policy education framework for university teaching and learning. *International Journal of Educational Technology in Higher Education*. https://doi.org/10.1186/s41239-023-00408-3

Fukuyama, F. (2020). 30 years of world politics: What has changed? *Journal of Democracy, 31*, 11.

Garrett, T. (2008). Student-centered and teacher-centered classroom management: A case study of three elementary teachers. *The Journal of Classroom Interaction, 43*, 34–47.

Hambrock, H., & De Villiers, F. (2023). Proposing a seamless learning experience design (SLED) framework based on international perspectives of educators from five higher education institutions. *Electronic Journal of e-Learning, 21*(1), 52–68.

II–Economic, D. G. (2001). European economy: 2001 broad economic policyguidelinesStatistical annex. number 72/2001.

Kinshuk, & Graf, S. (2012). Ubiquitous learning. In N. M. Seel (Ed.), *Encyclopedia of the sciences of learning*. Springer. https://doi.org/10.1007/978-1-4419-1428-6_224

Lenhart, A., Simon, M., & Graziano, M. (2001). *The internet and education: Findings of the pew internet & American life project.*

Perry, G. L., & Tobin, J. (Eds.). (2010). *Economic events, ideas, and policies: The 1960s and after.* Rowman & Littlefield.

Peters, E. E. (2010). Shifting to a student-centered science classroom: An exploration of teacher and student changes in perceptions and practices. *Journal of Science Teacher Education, 21*, 329–349.

Ramani, S. (2015). The internet and education in the developing world-hopes and reality. *Smart Learning Environments, 2*(1), 1–16.

Renard, L. (2023). *19-classroom-seating-arrangements-fit-for-your-teaching*. Retrieved from https://www.bookwidgets.com/blog/2019/12/19-classroom-seating-arrangements-fit-for-your-teaching

Sabah, S., & Du, X. (2018). University faculty's perceptions and practices of student centered learning in Qatar: Alignment or gap? *Journal of Applied Research in Higher Education, 10*(4), 514–533.

Swanson, E. (2014). August 4, 1964 50 years later. *The Choral Journal, 55*(1), 50.

Williamson, B. (2023). The social life of AI in education. *International Journal of Artificial Intelligence in Education*, 1–8. https://doi.org/10.1007/s40593-023-00342-5

Presentation

Developing a virtual culture in universities will require imaginative and creative implementation, as well as open leadership and an innovative mindset. Learning technology cannot be seen as a mere utility, but as an academic opportunity. Pedagogical design, multimedia production and data analysis are vital.

The evolution of the digital world, or a new technological breakthrough, or even a drastic change in the demands of the education market – the "educational technology phenomenon" – would have caused the gradual change we are witnessing. But not a virus. COVID-19 has redirected and amplified the concerns and actions of universities around the world, reshaping and challenging their interests in order to guarantee operational continuity in the short term, while ensuring institutional viability in the long term.

In this scenario, the behavioral changes that have emerged in the younger generations have inevitably led to the reformulation of learning itself. Consequently, teaching methodologies are being updated to keep up with the technological evolution of learning and, above all, to train future professionals for the job market and their professional careers.

In this context, there has been a digital transformation in higher education with the use of information and communication technologies. This transformation has led to the use of different technologies, including artificial intelligence (AI). The use of AI in higher education can provide personalized education through the automation of administrative teaching tasks, software programs that help detect topics that need reinforcement in the classroom, guidance and support for students outside the classroom and the intelligent use of data to teach and support students. In addition to providing personalized teaching using augmented reality, adaptive platforms, intelligent tutoring systems, chatbots, adaptive learning, computer-assisted instruction, direct feedback, Moocs and Robotics. So the questions in this ebook are:

How will the digital transformation take place in higher education? How can digital transformation be implemented in higher education? What are the links between Artificial Intelligence and the COVID-19 pandemic in the transformation of higher education? How can universities promote actions to implement digital transformation? What sustainable educational technologies can be used in the teaching-learning process for students? How can Blockchain technology and artificial intelligence be applied in higher education? How the metaverse can be applied in virtual learning environments in higher education? This book will present some situations in order to improve, enhance and qualify sustainable education through digital technologies.

The first chapter, entitled "Digital Transformations: Artificial Intelligence in Higher Education", by the authors Andreia de Bem Machado, João Rodrigues dos Santos, António Sacavém and Maria Jose Sousa, explains the applicability of artificial intelligence in higher education. Its aim is to map the digital transformation and the application of artificial intelligence in higher education. A bibliometric survey was carried out on the Web of Science database. The findings show that artificial intelligence can be applied to higher education through a smart contract that encourages collaboration between students and participation in extracurricular activities, as well as increasing machine reliability by creating a chain of information. In addition, it has been decided that this technology will help train a generation prepared to cooperate more and adopt sustainable attitudes to solve difficult everyday problems in the twenty-first century.

The second chapter entitled "Educational Institutions: Artificial Intelligence and Blockchain in Higher Education", by the authors Andreia de Bem Machado, Maria José Sousa, Gertrudes Aparecida Dandolini and Leonor Domingos aims to map, through a scoping assessment, how Artificial Intelligence and Blockchain can be used in education as a technology for higher education. It concludes that these technologies will contribute to the development of a generation prepared to work more co-operatively and adopt sustainable attitudes in local and global contexts to deal with the complicated issues of modern life.

The third chapter, entitled "Digital Transformation in Higher Education Institutions: A Case Study at Polytechnic University of Tomar", by the authors Célio Gonçalo Marques, Lígia Mateus and Inês Araújo, aims to report on a case study of the creation of digital educational content for the Polytechnic University of Tomar (IPT) using an agile learning approach. The process involves several stages, including planning, course design, content production, course creation on the project platform and course launch. This streamlined process simplifies content production and allows more teachers to integrate interactive learning materials into their teaching practice.

The fourth chapter, "Federal Public Policy and Digital Transformation in Higher Education", by Welinton Baxto, describes federal public policy and digital transformation in higher education. The findings point to the application of information and communication technologies in order to diversify access to formal education in the different regions of the country. The conclusion is that the federal government's guidelines are aimed at improving the teaching and learning process in higher education. However, it is in the process of fulfilling goal 12 of the National Education Plan, which refers to "democratizing access to higher education, with inclusion and quality". On the other hand, 9 years after the approval of the National Education Plan, new discussions are being held about the scope or otherwise of Law 13.005 of June 25, 2014.

The fifth chapter, "Augmenting Open and Distance Learning Through Gamification", by authors S K Pulist and Ramesh Chander Sharma, aims to investigate the pedagogy of gamification and how the gamification design framework can support active learning, performance monitoring and instant feedback. It concludes that gamification has been used in industry to increase productivity and acquire skills. And the instructional approach and nature of open and distance education study materials cater for the independent learner.

In the sixth chapter, entitled "Exploring the Potential of Gamification in Omani Higher Education Institutions: A Case Study of UTAS Salalah, Oman", by author Dr. Sangeeta Tripathi, the research examines the use of gamification in higher education institutions in Oman, with a specific focus on the University of Technology and Applied Sciences (UTAS) in Salalah. The results indicate that gamification has positively impacted student engagement, motivation and learning outcomes in a number of areas; however, there are some challenges faced by faculty members during the implementation of gamification, such as the need for training and support.

The seventh chapter, "Usages of ChatGPT in Higher Education: A Study with Graduate Students", by the authors Rogério Costa, Adriana Lage Costa and Ana Amélia Carvalho, aims to investigate how students use ChatGPT and their perspective on its functionality. A sample of 375 postgraduate students took part in the research, 76.5% of whom were female and 23.5% male. Most of the participants were master's students, while 22.4% are enrolled in a doctoral program. More than half of them (59.7%) reported using ChatGPT mainly in the academic context, followed by the social and professional context.

The eighth chapter by Paolo Biancone and Federico Chmet, entitled "Role of ChatGPT in Accounting", aims to examine the use of ChatGPT in accounting, highlighting its benefits and challenges. It explores how ChatGPT can simplify tasks ranging from data entry to preparing financial reports, while acknowledging data security and system integration concerns. It also highlights the evolving role of accounting professionals who can take advantage of ChatGPT for more in-depth analysis and strategic advice. Despite the fact that there are currently limitations in educational and professional contexts due to errors and limitations in thinking, this chapter highlights the continuous evolution of technology.

The ninth chapter, "A Review ChatGPT and the Future of Education," by the authors Andreia de Bem Machado, Antonio Pesqueria and Maria Jose Sousa, aims to map the future of education through ChatGPT in the context of a bibliometric review. It concludes that ChatGPT, which is used in various socio-economic areas, has been found. In the future of education, teachers and students will be able to use it. This ChatGPT can be used for various things, such as instruction and training, entertainment, question prediction, appointment setting and code debugging.

The tenth chapter, "Neolearning as a Methodology that Advances from Experiential and Expansive Learning Theories to Practice", by the authors Graziela Grando Bresolin and Patricia de Sá Freire, aims to analyze neolearning as a methodology that advances from experiential and expansive learning theories to practice. The results show student satisfaction with the application of the methodology in class, which provides greater motivation and applicability of the content in class, and teacher training to innovate in the classroom.

In the eleventh chapter, "Adopting Blockchain for Educational Qualifications in Italy: The Experience of the University of Turin", by the authors Katia Presutti and Francesca Natale, aims to provide an overview of blockchain in the Italian education system, addressing the main issues of stakeholders.

The twelfth chapter, entitled "Competences and Soft Skills in Higher Education: Innovation in the Assessment of University Students' Learning Outcomes", by Francesca Di Virgilio, presents the TECO (TEst on Competences) project, which focuses on meeting the needs of emerging sectors, innovative sectors and the promotion of student-centered teaching, accompanied by the analysis of learning outcomes, to accompany Italian higher education's bid for a sustainable and inclusive economic recovery.

Digital technologies and their applications have drastically transformed the world economy and global companies. The context also indicates that the growing market competition requires higher education institutions (HEIs) to transform their management structures and strategies. It is possible to see that this scenario of change has caused some traditional HEIs to focus their efforts and analyses on themselves, assuming that cost-cutting actions, reorganizing structures and digitizing classes are enough.

Contents

Digital Transformations: Artificial Intelligence in Higher Education

Andreia de Bem Machado, João Rodrigues dos Santos, António Sacavém, and Maria José Sousa

1 Introduction

The world of today is undergoing rapid change, making "invention" a crucial life skill. In this context, it is considered that the many socio-ecological issues that have arisen in the digital age have presented unique difficulties for society and call for creative solutions across all facets of society. This makes artificial intelligence an interdisciplinary field that focuses on the study, creation, and application of tools and methods for simulating, enhancing, and extending human intelligence (Zhong et al., 2018). It is possible that mankind is about to experience a new socio-technical paradigm that will have an unparalleled impact on world civilization and present prospects for transformation in many traditional economic areas.

The Dartmouth Society first introduced the idea of artificial intelligence in 1956 (Dodigovic, 2007), and since then, it has been used in a growing number of spheres of our life, including the setting of education. Building intelligent machines and artifacts, or at least ones that behave intelligently, is the goal of artificial

A. de Bem Machado (✉)
Federal University of Santa Catarina, Florianopolis, Brazil
e-mail: andreia.bem@ufsc.br

J. R. dos Santos
ISEC Lisboa – Instituto Superior de Educação e Ciências, Lisboa, Portugal
e-mail: joaopaulo.santos@iseclisboa.pt

A. Sacavém
Universidade Europeia, Lisboa, Portugal
e-mail: antonio.sacavem@universidadeeuropeia.pt

M. J. Sousa
ISCTE – University Institute of Lisbon, Lisboa, Portugal
e-mail: maria.jose.sousa@iscte-iul.pt

A. de Bem Machado et al. (eds.), *Digital Transformation in Higher Education Institutions*, EAI/Springer Innovations in Communication and Computing,
https://doi.org/10.1007/978-3-031-52296-3_1

intelligence. Artificial intelligence technology is a crucial branch of computer science that focuses on the study and advancement of technical disciplines used to mimic, amplify, and extend human intelligence. Artificial intelligence has ushered in an explosive era in recent years because of the great strides made in machine processing and machine learning as well as the exponential growth of data. Artificial intelligence has the potential to radically transform several sectors, including security, banking, retail, transportation, and education, thanks to its advantages in analysis, prediction, judgment, and decision-making.

Therefore, the objective of this research is to map digital transformation and artificial intelligence in higher education. To achieve this goal, we performed a bibliometric review. This research is divided into six parts: The first is the introduction, the second explains the digital transformation theme, the third addresses artificial intelligence, the fourth the methodology adopted, the fifth part discusses digital transformations: artificial intelligence in higher education, and, finally, the final considerations of the authors.

2 Digital Transformation

Digital transformation refers to the integration of digital technology into all organizational areas, changing how operations are conducted and the way value is delivered to customers (Berman, 2012). It is not just a buzzword or about updating the technology within an organization, but about reshaping the organization's culture and operations to embrace continuous change (Bounfour, 2016). Indeed, digital transformation entails a fundamental shift in how businesses operate and deliver value since is driven by a multitude of factors, namely: increasingly pervasive and interconnected digital technology, changes in customer behavior and expectations, and the emergence of innovative business models (Matt et al., 2015). A holistic and strategic approach is critical for a successful transformation, which considers technology, data, process, and organizational change. In this scenario, one of the most comprehensive and balanced interpretations views digital transformations as a progressive process that utilizes digital technologies and abilities to create value-driven business models, enhance business practices and operations, and improve service delivery (Morakanyane et al., 2017).

The digital age has created a profound impact on organizations. Those failing to adapt and evolve face the risk of falling behind in the competitive landscape (Kane et al., 2015). Nowadays, organizations need to harness the power of digital transformation to improve business processes, corporate culture, and customer experiences to meet organizational change and market requirements (Westerman et al., 2011). The significance of digital transformation reaches into the sphere of higher education, where it can notably improve the learning experience by making it more accessible, engaging, and efficient. By adopting digital transformation, higher education institutions will be more equipped to cater to the evolving demands of students and the wider academic society. One of the universities purposes has changed to training future managers to tackle issues and look for solutions, including information literacy (Akour & Alenezi, 2022). Therefore, institutions are leveraging technology to enhance learning experiences, streamline administrative tasks, and promote innovative research practices (Daniel, 2015).

If there is a comprehensive understanding of the skills required for the future and how universities are training students to acquire these skills, it would lead to enhancements in teaching, learning, and university functions, benefiting all involved parties. This understanding is particularly crucial given the rapid pace of workplace transformation (Alenezi, 2021). In the face of evolving technology, it is vital for higher education institutions to navigate the uncertainty in establishing robust and adaptable educational support systems that meet both educational goals and future employment needs. Universities remain key in preparing students with the necessary skills for success in the twenty-first-century workplace. However, the swift pace of technological advancement is posing challenges for these institutions in effectively planning, strategizing, and predicting a spectrum of educational goals (Akour & Alenezi, 2022). To continue playing a significant role in preparing students for future careers, they need to overcome these challenges. As a result of technological progress, educational institutions may find themselves needing to prioritize courses that focus on technical, advanced cognitive, knowledge-based, and digital skills, potentially overshadowing more holistic, cooperative, multidisciplinary, and cross-cultural learning methods. Some also argue that technology is reshaping the dynamics between students and institutions in terms of learning expectations and requirements. This has spurred the growth of competitive and alternative learning methods, fostering more flexible, comprehensive, and personalized learning experiences (Akour & Alenezi, 2022).

Digital transformation has made higher education more accessible and flexible. Innovative technologies like ubiquitous and pervasive computing, smart devices, along with the proliferation of massive open online courses and extensive use of social media, are driving the digital transformation, which is fundamentally altering the methods and accessibility of teaching, learning, and assessment (Bamiah et al., 2018). This flexibility is particularly beneficial for non-traditional students who may be balancing education with work or family responsibilities.

Digital resources like interactive simulations, virtual reality, and collaborative online platforms can significantly enhance student engagement and collaboration (Bonk & Khoo, 2014) since these tools can make learning more interactive and fun, thereby improving student motivation and learning outcomes. Additionally, given the pervasive presence of social media in a student's daily life, it can serve as a potent instrument in facilitating teaching, peer interaction, and communication with instructors. The utilization of social media promotes student involvement, peer interaction, and student–teacher engagement (Akour & Alenezi, 2022). Creating strategies to encourage, foster, and stimulate the use of social media in a virtual learning setting, along with offering efficient implementation and oversight for educators adopting social media more extensively, could be beneficial for both students and teachers, leading to superior results in educational institutions (Bozanta & Mardikyan, 2017). Furthermore, adaptive learning systems and learning analytics can provide real-time feedback to students and instructors, enabling them to effectively adjust to the learning process. Moreover, digital transformation enables personalized learning, because teaching is tailored to each student's unique needs and distinctive learning pace (Becker et al., 2018).

Several studies have scrutinized the efficiency of students' engagement with e-learning platforms, focusing on their dependence on, acceptance of, and incorporation of technology into academic tasks. The patterns of study differ, and students' perceptions of the platforms' usefulness and user-friendliness influence their willingness to use them effectively (Moreno et al., 2017). As a result, educational institutions should promote and facilitate the incorporation of e-learning platform features into instructional activities (Akour & Alenezi, 2022).

Digital transformation has the potential to boost research abilities in higher education. Tools such as digital databases, data analytics, and cooperative research platforms can streamline, improve the research process (Borgman, 2017), and assist researchers in swiftly and precisely analyzing extensive datasets, resulting in more substantial research conclusions (Dwivedi et al., 2020). Furthermore, digital transformation can promote international cooperation, enabling researchers worldwide to collaborate more effectively, and it can also facilitate instantaneous data gathering and examination, resulting in more precise and prompt research findings (Mohamed Hashim et al., 2021). Finally, digital transformation could offer a heuristic model to foresee potential risks and opportunities for researchers, adding to the discussion about the implications of digital transformation on public higher education and research (Barzman et al., 2021).

Despite its necessity and outstanding opportunities, digital transformation is not an easy task. Many organizations face significant challenges, such as resistance to change, lack of digital skills, and the complexity of integrating new technologies into existing infrastructures (Verhoef et al., 2021). Digital transformation in higher education also poses challenges. These include the resistance to change, the need for upskilling staff, the financial cost of implementing new technologies, and concerns over data privacy and security (Selwyn, 2013). According to Alenezi (2021), the digital transformation challenges in higher education institutions are: prioritization; decentralized decision-making; human resistance to change; gaps in digital tech talent, and narrow view of ROI. Universities often concentrate on pressing issues at the expense of significant ones, prioritizing immediate outcomes while delaying crucial investments in capacity building that require time. It's argued that the approach to planned digitization is neither particularly innovative nor adequately funded (Rodríguez-Abitia & Bribiesca-Correa, 2021). Decentralized decision-making can lead to postponements in executing extensive decisions and projects. Conversely, IT systems governed by centralized control and decision-making tend to be more effective. This is a crucial method to balance the magnitude and pace of coordinated decision-making. For successful outcomes, it's necessary to establish a flexible dual-handling organizational model for digital transformation, which can optimally balance segregation and integration over time (Smith & Beretta, 2021). There's a consensus that the main challenge faced by higher education institutions in adapting to digital transformation lies in adjusting to new teaching methodologies and learning environments and models. Therefore, resistance to embracing digital transformation can pose a significant hurdle. Academic professions are often considered among the most stable jobs globally. If change poses a threat to job security, it's likely to be met with opposition (Rodríguez-Abitia & Bribiesca-Correa, 2021); therefore, it is crucial to establish guidelines for educational professionals to enlighten them about the benefits of technology and alleviate their concerns about job security (Alenezi, 2021). The primary obstacle to successfully integrating digital technologies into the higher education system is the generational divide between students, who are inherently familiar with these technologies, and the faculty, who must adjust and learn to use them. Therefore, due to these generational differences between students and faculty adopting technology, it is essential to advocate for policies that facilitate both the infrastructure and innovative learning environments to keep up with evolving scholarly trends. Moreover, significant emphasis will need to be placed on faculty training (Alenezi, 2021). Finally, it is observed that higher education institutions often make mistakes in determining the business case and return on investment (ROI) for their investments. Business cases for fundamental digital skills should consider actions that evolve slowly or are challenging to quantify. These may include saving faculty time and costs through more efficient operations and automation, as well as improved enrollment, retention, and timely graduation rates resulting from an enhanced student experience (Alenezi, 2021).

Notwithstanding these challenges, with appropriate tools and creativity, challenges can be transformed into opportunities. Artificial intelligence (AI) may help in the endeavor, since encompass forecasting learning progress, performance, satisfaction, suggesting resources, automated evaluation, and enhancing the learning experience. Moreover, AI-enable suggestions of high quality tailored to student attributes, enhancement of students' academic achievements, and increased online engagement and participation (Ouyang et al., 2022).

3 Artificial Intelligence

John McCarthy was the first to define artificial intelligence in a two-month workshop at Dartmouth College in the US, even though several definitions have surfaced in recent years. Artificial intelligence (AI) is a branch of study that focuses on artificially imitating human intelligence's cognitive capacities to build computer programs or other devices that can carry out tasks that are traditionally handled by people. According to Norvig and Russell (2013, p. 24), the phrase "artificial intelligence" is used "when a machine mimics the cognitive functions that humans associate with other human minds, such as learning and problem-solving." McCarthy used the phrase "artificial intelligence" for the first time in 1956 in the workshop proposal (Russel & Norvig, 2010). They explicitly state that AI is the science and engineering of creating intelligent machines, particularly intelligent computer programs, in their definition of the term. While it is related to the related job of utilizing computers to study human intelligence, AI need not be constrained to techniques that can be observed biologically.

AI will function under the presumption that, in theory, every facet of learning or any other characteristic of intelligence can be sufficiently correctly characterized for a machine to replicate. There will be an effort to figure out how to make machines understand language, create abstractions and notions, solve human-only problems, and advance (Zawacki-Richter et al., 2019).

Computers that do cognitive activities, generally similar to those performed by the human mind, especially learning and problem-solving, are how Aldredge et al. (2021) define AI. Additionally, they clarify that AI does not refer to a specific technology. It serves as a catch-all phrase for a variety of technologies and approaches, including genetic algorithms, machine learning, natural language processing, data mining, and neural networks.

Another way to define AI is as a group of models, techniques, and technologies (such as knowledge searching, reasoning, and representation, decision mechanisms, perception, and planning, as well as natural language processing, uncertainty handling, and machine learning) that, either individually or collectively, can resolve issues involving logical reasoning (Woods et al., 2022). As a result, the term "artificial intelligence" (AI) has a wide meaning that includes several capabilities, including knowledge, reasoning, problem-solving, perception, learning, planning, and the ability to handle and move objects (Zawacki-Richter et al., 2019). The presence of AI in various situations, including the educational context – which has progressively integrated technology instruments into the pedagogical process – is considered in this scenario.

Learning has changed as a result of the behavioral shifts that have occurred in younger generations. To prepare future professionals for both the labor market and professional vocations, teaching approaches are being changed to keep up with the technological advancements in education (Bucea-Manea-oniş et al., 2021). The ability of machines, i.e., computers, to accomplish goals through learning based on prior experiences linked with people has been termed as AI in this situation. AI is a system's capacity to use input data to learn in order to accomplish a specific goal. Its acceptance extends beyond the field of computer science and touches on fields like information science, psychology, linguistics, and mathematics, among others. To create integrative and adaptable settings that enhance and supplement conventional educational approaches, artificial intelligence (AI) integrates techniques and findings from various scientific disciplines (Bucea-Manea-oniş et al., 2021).

In this context, artificial intelligence (AI) can be applied in higher education for individualized learning through the automation of administrative teaching tasks, software that supports the detection of topics that need reinforcement in the classroom, the guidance and support of students outside of the classroom, and the intelligent use of data to teach and support students. AI can be used in education in a variety of ways beyond than personalized instruction, including robotics, augmented reality, adaptive platforms, intelligent tutoring systems, chatbots, adaptive learning, computer aided instruction, direct feedback, and MOOCs (Kuleto et al., 2022).

In the context of education, research demonstrates that online courses, MOOCs, augmented and virtual reality, game-based learning, and collaborative work are all quantitatively linked to student learning results. Thus, Augmented Reality (AR), a technological advancement in education that has the capacity to draw students' attention to an engaging and enjoyable learning environment, can be used to demonstrate the relevance of AI to higher education (Ngabo et al., 2021). Therefore, studies on the use of AI in higher education focus on management systems and student learning opportunities. Machine learning (ML) techniques are used in conjunction with natural language processing (NLP) to identify patterns in huge datasets (human-written sources). By looking up posts on social networks, it can be used to analyze how buyers feel about specific goods or services (Aldredge et al., 2021).

Intelligent tutoring systems are a component of the AI story that enables the personalization of education by encouraging interactive learning through the creation of tutoring programs (Wang et al., 2021). These are a few of the cutting-edge technical options for expanding inclusive educational learning. While distributed intelligent teaching systems promote student collaboration through software programs that support and encourage interactions, intelligent tutors act as a guide for student learning, detecting student progress in learning based on content knowledge and the student's personal characteristics (Bucea-Manea-oniş et al., 2022).

Chatbots are a different AI-related technology that is used in higher education. In order to engage students individually, collect candidate data, and construct virtual teaching assistants utilizing this data, higher education institutions have developed chatbots (Kuleto et al., 2021). A chatbot is a piece of software that mimics text- or voice-based conversations with people. It can comprehend spoken language and pick up new vocabulary through usage. Watson Assistant (IBM), Siri (Apple), Alexa (Amazon), and Google Assistant (Google) are a few instances of well-known virtual assistants.

4 Methodology

To map digital transformation and artificial intelligence in higher education, we chose to use bibliometric analysis. Considered a science, bibliometrics involves the quantitative and statistical study of the production, publication, use, and dissemination of scientific knowledge from bibliographic databases (Gutiérrez-Salcedo et al., 2018).

Through analytics and statistical tools, it allows to evaluate, verify, identify, and monitor the progress of fields of study through the classification of data such as citations and collaborations, author affiliations, keywords, topics discussed, relevant topics, and methods used (Ali et al., 2016).

The study was divided into three independent phases for the bibliometric analysis: planning, data collecting, and findings. These steps came together to respond to the study's central query: How can artificial intelligence technology be used in higher education?

Planning lasted from September through May 2023, when the research was carried out. Due to the quantity of documents deemed sufficient in the research bases in the database selected for this research, some criteria were defined in this phase as the limitation of the search in electronic databases, which in this research's case was the Web of Science (WOS). Due to the WoS database's importance in academia and its interdisciplinary nature, which is the focus of this area of research, it was specified in the planning scope as pertinent to the research domain. And also due to the fact that it is one of the largest databases of summaries and bibliographic references of peer-reviewed scientific literature and its constant updating (de Bem Machado et al., 2022). The terms used in the WOS database search were: "Digital Transformations" and "Higher Education."

Since a notion depends on the context to which it is related, it is believed that the variations of the search terms are offered, in a bigger context, inside the same proposition. Finally, it was chosen to just use the terms defined in the "title, abstract, and keyword" sections when planning the search, excluding any additional criteria that would have limited the results, such as time or language. Furthermore, documents of the "article" type were used as the inclusion criterion.

From the research planning, the data collection, carried out on 30 May 2023, retrieved a total of 13, indexed papers, which pointed registration from 2019, the first publication, until 2023.

4.1 Results of Bibliometry

As a result of this collection, it was identified that the 13 papers identified in the search were written by 41 authors, linked to 11 different countries. A total of 86 keywords were used. Table 1 shows the results of this data collection in a general bibliometric analysis.

Table 1 Bibliometric data

Description	Results
MAIN INFORMATION ABOUT DATA	
Timespan	2019:2023
Documents	13
DOCUMENT CONTENTS	
Author's Keywords (DE)	86
AUTHORS	
Authors	41
AUTHORS COLLABORATION	
Single-authored docs	1
Co-Authors per Doc	3,15
International co-authorships %	23,08
DOCUMENT TYPES	
Article	13

The eligible articles in the Web of Science database were published in the period from 2019 to 2023. Being that in the year 2019, there were three publications. In the year 2020, three papers were published. In the year 2021, there were two publications. And in the years 2022, there were three published papers and in the year 2023 there were two publications until the time of data collection of the research, according to Fig. 1.

From the 13 papers, one can observe a varied list of authors, institutions and countries that stand out in research on digital transformation, AI and higher education. When analyzing the 11 countries with the highest number of citations in the area, Russia stands out with 20% of total publications. In second place, stands out India with 17% of citations, as shown in Fig. 2.

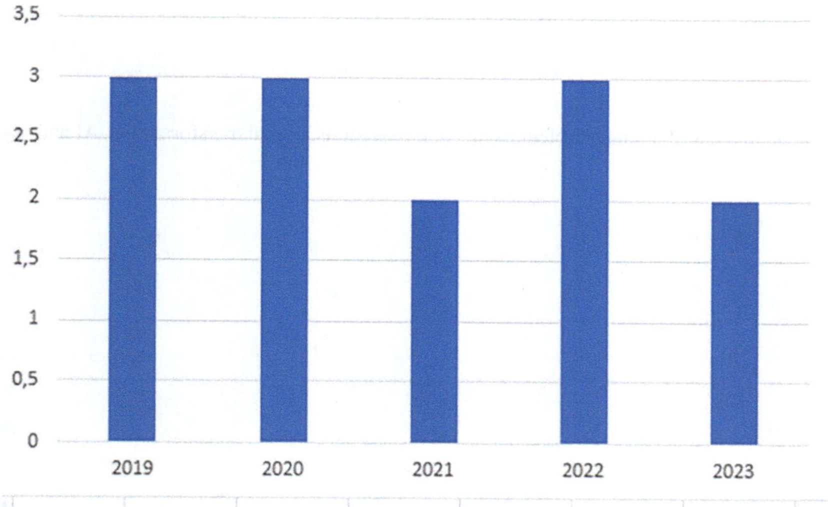

Fig. 1 Temporal distribution

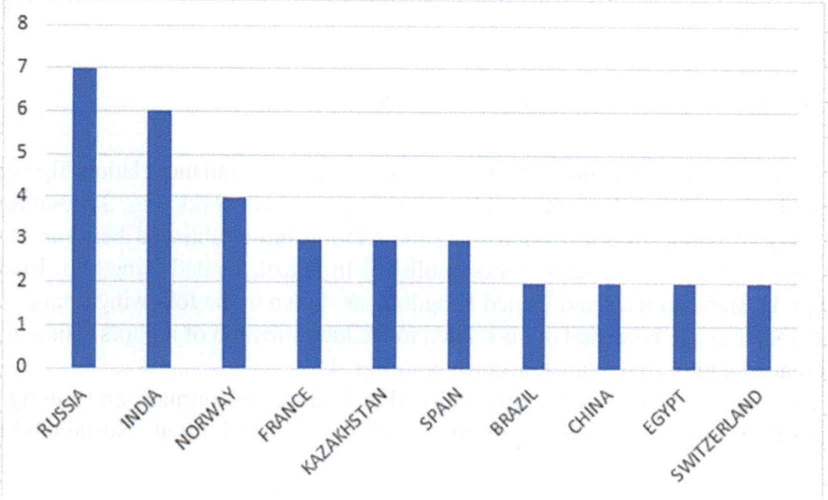

Fig. 2 Distribution of the papers by country

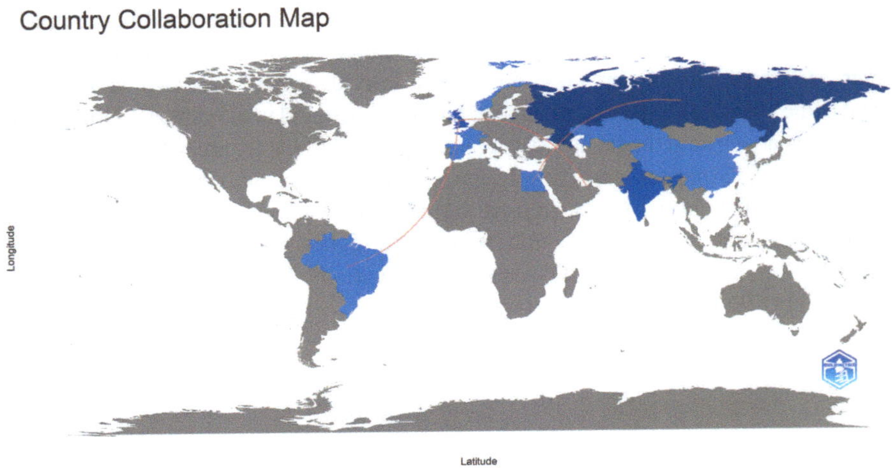

Fig. 3 Spatial distribution and relationships of publications, digital transformation, AI and higher education

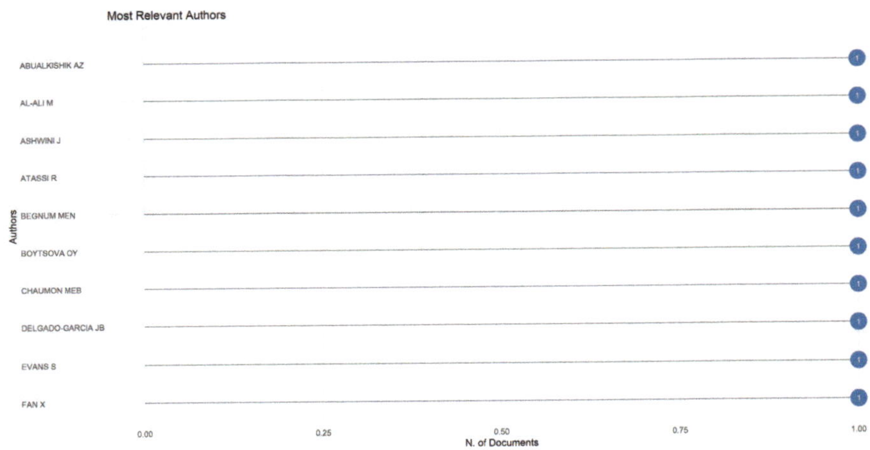

Fig. 4 Authors with publications on the subject of the search

Figure 3 presents the intensity of publication by country and the relationship established between them, through citations between published works, Fig. 3 presents the intensity of publication by country and the relationship established between them, through of citations between works published in Brazil, United Kingdom, Russia, Egypt, U Arab Emirates and United Kingdom, as shown in the following image:

Another analysis carried out is related to the identification of authors. There is no referenced author in this area, according to Fig. 4.

The documents analyzed were published in 12 different journals, and among the total of 13 studies, 2 (19%) were published in only one journal "Sustainability,"

Table 2 Scientific sources with most publications

Sources	Number of publications
SUSTAINABILITY	2
EDUCATION AND INFORMATION TECHNOLOGIES	1
FRONTIERS IN PSYCHOLOGY	1
INTERACTING WITH COMPUTERS	1
INTERNATIONAL JOURNAL OF ADVANCED COMPUTER SCIENCE AND APPLICATIONS	1
INTERNATIONAL JOURNAL OF EARLY CHILDHOOD SPECIAL EDUCATION	1
INTERNATIONAL JOURNAL OF EDUCATION AND INFORMATION TECHNOLOGIES	1
INTERNATIONAL JOURNAL OF EMERGING TECHNOLOGIES IN LEARNING	1
JOURNAL OF CLEANER PRODUCTION	1
PSYCHOLOGIE DU TRAVAIL ET DES ORGANISATIONS	1

Fig. 5 Tag cloud

according to Table 2, which presents the 12 scientific sources with the most publications on the theme "Digital transformation, AI and higher education."

From the bibliometric analysis, based on the work group retrieved, on the basis of 86 keywords indicated by the authors, digital transformation gained prominence with 5.education with three occurrences, according to Fig. 5.

Also highlighted is "digital transformation," from which we conclude that this is an area of application of AI technology in higher education that has been extensively explored in the literature.

5 Digital Transformations: Artificial Intelligence in Higher Education

In the twenty-first century, artificial intelligence plays a crucial role in various aspects of society, covering areas such as technology, health, economy and finance, industry, transportation, education, among others.

Artificial intelligence is used to automate repetitive processes and tasks, freeing up human beings to focus on more complex and creative/innovative activities. This necessarily results in greater efficiency and productivity.

For example, in some Chinese schools, an artificial intelligence device is used that lets teachers know in real time when any student is not concentrating. “Headbands measure each student’s concentration level and then send information directly to the teacher and parents” (Panchal et al., 2020, p. 1).

Another key valence of artificial intelligence is facilitation of data analysis, enabling rapid and accurate analysis of large volumes of data, identifying patterns, trends, and hidden insights. This is especially useful in areas such as finance, marketing, market forecasting, and strategic decision-making.

“Personal digital assistants like Siri from Apple, Alexa from Amazon, Microsoft’s Cortana or Assistant from Google are at the forefront of technology of voice recognition and artificial intelligence” (Smutny & Schreiberova, 2020). These virtual “assistants” are able to understand voice commands, provide information, perform tasks, and even hold conversations.

Medicine and healthcare, autonomizing motor vehicles, or personalizing recommendations on the various platforms of the World Wide Web are other examples of how artificial intelligence is impacting our lives today.

For example, Price II, (2017) indicate that:

> Medicine, like many other fields, is experiencing a confluence of two recent developments: the rise of big data, and the growth of sophisticated machine learning/artificial-intelligence techniques that can be used to find complex patterns in those data. (p. 2)

With this possibility, it is to healthcare professionals to make more accurate and personalized decisions.

Also with artificial intelligence, it is possible for motor vehicles and drones, for example, to “learn” to make decisions based on data and sensors, making transportation and travel safer, more efficient, and more effective.

Now in the field of automotive transportation, “developments in the IoV [Internet of Vehicles] domain are faster than ever, due to advances in Internet technologies and vehicle equipment” (Mandal & Sarkar, 2020, p. 5).

Still about the IoV, the same authors mention that one of the layers of the architecture is “Perception.” According to them (2023):

> This layer represents the interaction between the vehicle and its environment. It uses the devices present inside the vehicles, such as sensors, actuators, personal devices and those installed along the road, such as MSW, in order to collect relevant information to be used in the vehicle decisions. (p. 3)

Also, any World Wide Web user these days benefits from artificial intelligence by receiving personalized recommendations on streaming services, e-commerce platforms, and social media. This ability to understand individual preferences improves the user experience and boosts sales.

In the field of education, for example, Smutny and Schreiberova (2020) report that today "conversational pedagogical agents use artificial intelligence techniques to enhance and personalize automation in teaching" (p. 2). "Artificial intelligence can enhance students' learning experience and help them achieve more in less time" (Geerling et al., 2023, p. 8).

These are just a few examples of the broad spectrum of applications of artificial intelligence in the twenty-first century. Since the technology is sequential and incremental, as it advances, its role will expand much further, hopefully positively affecting various industries and areas of human life. With the technological dynamics of artificial intelligence, the future is unpredictable. One thing is certain: "the past ten years have been the golden decade of AI, but meaningful AI has just begun" (Hu, 2023).

Therefore, generically, one could say that artificial intelligence is a field of computer science that focuses on developing systems capable of performing tasks that would, under normal conditions, require human intelligence.

According to Zeide (2019):

> Today's artificial intelligence is based on machine learning. It is about finding patterns in seas of data-correlations that would not be immediately intuitive or comprehensible to humans-and then using those patterns to make decisions. With "predictive analytics," data scientists use past patterns to guess what is likely to happen, or how an individual will act, in the future.

Currently, because of its pervasive nature, artificial intelligence is applied in a variety of industries, such as medicine, transportation, finance, manufacturing, higher education, and many others.

Surden (2019), citing Ting et al., (2018), clarifies that the "most successful artificial technological approaches fall into two broad categories: (1) machine learning and (2) logical rules and knowledge representation" (p. 1310).

Regarding the use of artificial intelligence in education, Geerling et al. (2023) point out that "the advent of artificial intelligence in education is a reality that cannot be ignored, and it is time to embrace the new era with innovative and effective assessment strategies" (p. 9).

Specifically, in relation to higher education, artificial intelligence is playing an increasingly significant role, bringing benefits and advancements in various areas of focus in academia.

Baidoo-Anu and Owusu (2023), citing Petersen (2021), note that "the twenty-first century has experienced a rapidly changing landscape in educational practices largely due to advancement in technology (such as artificial intelligence)" (p. 4).

In this context of technological disruption, the value of learning personalization stands out. Based on artificial intelligence algorithms, educational institutions can personalize the learning experience for each student, adapting content, teaching speed, and assessment strategies to individual needs. For example, in relation to assessment, Petersen (2021), emphasizes that:

> As the level and complexity of student learning have increased, technology is playing a critical role in improving assessment. Smart 'adaptive tests' adjust the difficulty of assessment items as students' progress through them for more precise measurement (p. 6)

In another dimension of learning, artificial intelligence systems can provide virtual tutoring to students, assisting them with answers to questions, providing guidance on study material, and offering personalized feedback on their academic performance. In the same vein, artificial intelligence, in addition to being used to facilitate the teaching–learning process, can also be instrumental in research, helping to analyze large sets of academic data and identify trends and patterns in performance.

For example, identifying factors affecting student success is a very relevant area of focus for artificial intelligence, as it can help institutions identify students at risk of dropping out and offer timely interventions. In this sense, Panchal et al. (2020) highlight that "schools in China have adopted AI . They have facial recognition technology. The systems are [...] used to record student attendance and handle enrollments" (p. 1).

The systems mentioned above examine seven distinct expressions: (1) Neural; (2) Happy; (3) Scared; (4) Disappointed; (5) Angry; (6) Surprised. (p. 3) "This all helps to tell how much students are engaging in classroom activities" (Panchal et al., 2020, p. 4).

Finally, higher education institutions today can also take advantage of artificial intelligence for the benefit of students. Chatbots, also based on artificial intelligence, can provide guidance and support to students during the application and enrollment process, making the process easier and faster. In fact, in the context of artificial intelligence and the use of chatbots as educational aids, Sinha (2020) highlights the following: "[artificial intelligence] can help in making learning more adaptive and personalized. Teachers are finding AI driven Chatbots very useful at different stages to enhance the student experience" (p. 10). "Chatbots have been used across a wide range of domains, including marketing, customer service, technical support, as well as education and training" (Smutny & Schreiberova, 2020, p. 1).

Indeed, regardless of the areas of focus related to higher education, it is certain that artificial intelligence will accelerate the ongoing digital transformation and therefore also the future of higher education generically considered. According to Marks and AL-Ali (2022):

> Digital transformation can support higher education to shift from wondering about the future into predicting, forecasting, and shaping the future; making proactive and informed decisions and taking action based on that information. (p. 507)

Next, two successful real cases of applying artificial intelligence to higher education are presented: the case of Georgia State University; and the case of Stanford University.

A very interesting real-life example of artificial intelligence in higher education is the case of Georgia State University in the United States of America. The university used artificial intelligence to improve its graduation rate and reduce student dropout.

Through an artificial intelligence-based system, the university analyzed academic data from thousands of students over several years. "The university used a data-based approach to retain students of all racial, ethnic, and socioeconomic backgrounds and help them graduate" (Harvard Business Review, 2021). Based on these insights, the university developed customized interventions for at-risk students.

"After implementing AI, Georgia State University reported a 21.4% reduction in summer melt and 3.3% higher enrollment rate, resulting in $3 million of tuition revenue" (Baraishuk, 2021).

The successful case of Georgia State University is an excellent example of demonstrating how artificial intelligence can be used to analyze large volumes of data, identify patterns, and provide personalized interventions, helping educational institutions improve student performance and success.

Another real case of applying artificial intelligence in higher education is the use of technology to assist in the automated correction of essays and essay assessments. An excellent example in this regard is Stanford University, which has developed a system called Stanford Automated Essay Scoring (SAES). This system uses artificial intelligence algorithms to analyze and grade essays written by students. According to Song and Zhao (2013), "the model systematically classifies the quality of writing and can be applied in both academia and large industrial organizations to improve operational efficiency" (p. 1).

Ke & Ng (2019) stress that an "automated essay scoring (AES), the task of employing computer technology to score written text, is one of the most important educational applications of natural language processing" (p. 6300). The advantage of this approach is that automated correction allows for a faster and more consistent assessment process, reducing teacher workload. In addition, the instantaneous feedback provided by the system allows students to improve their writing skills more quickly over time.

The example of SAES and Stanford University perfectly illustrates how artificial intelligence can be applied in higher education to improve assessment processes, provide rapid feedback, and help students improve their writing skills.

The two cases described above are just two examples of how artificial intelligence can increase the quality and effectiveness of higher education in its various domains. Marks and AL-Ali (2022) state that "artificial Intelligence, [...] can enhance and change educational system practices, especially in a world where students are constantly interacting with technology in all other aspects of their everyday lives" (p. 507).

Broadly speaking, the use of artificial intelligence in higher education brings with it several significant virtues. Among the most important virtues of artificial intelligence today is the personalization of learning: Artificial intelligence allows teaching to be tailored to the individual needs of students. This includes identifying knowledge gaps and providing personalized content to improve the effectiveness of the teaching–learning process. This improvement in the effectiveness of the process is the result of a number of factors. First, the possibility for students to get immediate feedback on their performance.

Despite all the advantages listed, it is important to emphasize that artificial intelligence does not replace the role of educators in higher education. Artificial intelligence complements the teachers'skills and the other resources available, providing a more personalized and, therefore, more effective teaching–learning process.

Still, the application of artificial intelligence to higher education is not without concerns. On the contrary. For example, Geerling et al. (2023) emphasize that "the rise of artificial intelligence in higher education, specifically natural language models like ChatGPT, presents a new challenge to educators" (p. 7).

The major concern underlying the application of artificial intelligence to knowledge construction in general and higher education in particular is precisely the prevalence of the idea that human critical thinking is eventually no longer necessary.

In this context, Surden (2019) highlights that, for example, "in areas of law or legal practice that involve judgment, human cognition will likely be difficult to replace given the current state of AI technology" (p. 1330).

It will always be algorithms that are at issue when addressing the phenomenon of artificial intelligence. For this reason, there are several concerns that should be considered when using artificial intelligence in higher education.

One of the biggest challenges is the possible, and likely, algorithmic bias. For example, it is true that a tool like ChatGPT is "trained on a vast amount of text for its predictive algorithm, which gives it a significant advantage over its human counterparts" (Geerling et al., 2023, p. 9). But it is still algorithms we are talking about, devoid of awareness and critical thinking. And so, these artificial intelligence systems may reproduce and amplify biases present in the data used and "interpreted." This eventuality could lead to discriminatory or unfair outcomes, affecting, for example, access to education and educational equity.

According to Zeide (2019), "several elements need to be considered to ensure that the implementation of AI tools is optimal and equitable" (p. 9). Some of these elements are Procurement, Training, Oversight, Policies and Principles, and Participation.

"Despite its inherent limitations, it is a nearly undeniable fact that ChatGPT and other generative AI have come to stay and will continue revolutionizing the current educational system" (Baidoo-Anu & Owusu, 2023, p. 16).

Another very relevant concern has to do with ensuring data privacy and security. For example, in this regard, according to Travis et al. (2006), "in the United States, federal and state regulations prescribe stakeholder rights and obligations that must be satisfied by the requirements for software systems."

The use of artificial intelligence involves the collection and storage of large amounts of personal student data. Ensuring the protection of this data from unauthorized access and misuse is essential, but it can also pose an immense challenge. Giving an example, Ting et al., (2018), in this regard, states that:

> At the computer science department of Carnegie Mellon University have modeled the problem of compliance with privacy law as one of inputs (data) and outputs (certain restrictions on the scope and extent of permissible data sharing). (p. 18)

From another perspective, in some cases, artificial intelligence systems can be considered "black boxes" – absolutely hermetic, armored and impenetrable systems, figuratively speaking. "Unpacking what is occurring within AI systems is very difficult because they are dealing with so many variables at such a complex level" (Zeide, 2019, p. 7). This could create distrust and even make it difficult to provide justifications for the options chosen.

As mentioned earlier, full teacher replacement is also one of the major concerns underlying the use of artificial intelligence in higher education. The accentuation of this process may raise ethical questions. Ethics does not dispense with critical thinking. According to Zeide (2019), "ethical questions revolve around consequences in terms of different groups and subgroups, educational values, and how AI systems might alter those values" (p. 7). While automation of certain tasks is possible, it is important to recognize the irreplaceable value of educators in higher education. This value stems from the cognitive ability and critical thinking that are still being held exclusively by humans. "The vision of AI as involving thinking machines with abilities that meet or surpass human-level cognition – often referred to as Strong AI or Artificial General Intelligence (AGI) – is only aspirational" (Surden, 2019, p. 1309, citing Mills, 2018). While it may seem paradoxical, it is also a concern that humans, endowed with critical thinking, may become overly dependent on artificial intelligence and automated systems. This could create an irresolvable vulnerability should technical failures or interruptions in service provision occur. Reinforcing this idea, Surden (2019), in the context of law and the law, states that "there are many lawyering tasks that involve abstraction, conceptualization, and other cognitive tasks that current AI technology is not good at" (p. 1331).

Therefore, it is essential to maintain a balance between technology and human presence to ensure the continuity of operations inherent to the teaching–learning processes. Reinforcing this premise, Baidoo-Anu and Owusu (2023), call attention to the following:

> Despite the myriad of potential educational benefits, in its current state ChatGPT has been found to have several serious inherent limitations, such as generating wrong answers and making up articles that do not exist. (p. 15)

From the perspective of equity and info-exclusion, it is also important to emphasize that the use of artificial intelligence technologies in higher education can create access inequalities, especially in regions with limited resources or marginalized populations. It is necessary to ensure that all students have equal access to educational opportunities.

In practice, the adoption of artificial intelligence in higher education should presuppose the proactive implementation of appropriate safeguards, promoting transparency and ensuring both equity and inclusion in the educational process.

And ensuring equity should be precisely one of the major vectors to guide, in the future, the adoption of artificial intelligence in higher education.

Therefore, by avoiding info-exclusion and carefully moderating the use of artificial intelligence, balancing it with the indispensable human presence, the future of the use of artificial intelligence in higher education can only be promising, since it offers teachers, students and university staff a wide range of possibilities for increasing the effectiveness of the processes related to teaching–learning.

"Generative AI is a nascent but creative approach. It is one of the most successful ML [machine learning] frameworks in the evolution of deep learning over the past decade" (AI) (Hu, 2023).

From improvements in adaptive and personalized learning, through the enhancement of virtual tutoring and learning assistants, to, for example, virtual and augmented reality, predictive analytics and early detection of difficulties, intelligent collaboration, and the facilitation of global accessibility to education, the application of artificial intelligence to higher education will bring absolutely vital advantages for global development and economic growth based on increasingly specialized knowledge. It is true that "in order to generate products and services, an economy uses available technology to combine the factors of production" (Santos, 2020, p. 28). Technology will never cease to be a determining factor. Nevertheless, there is an important caveat here. Research in economics education suggests that the teaching–learning process in this area of knowledge can benefit greatly from activities that require human interaction and the use of critical thinking. Examples of these activities include the experiments, innovative assignments, classroom demonstrations, undergraduate research, the use of social media, the use of media and pop culture, case studies, and cooperative learning. (Ghent & Mateer, 2016).

Even for the example described above – of teaching economics – it is important to point out that the implementation of technologies in the educational field will require a careful balance between the use of artificial intelligence and human presence. Mills (2018) ironizes about this important aspect: "AI would expect a machine to perform functions that are now only seen in science fiction robots." Educators will continue to play a key role in higher education, guiding and supporting students on their educational paths, while artificial intelligence offers enhanced and personalized resources to enhance the learning experience.

Finally, one of the big issues for the near future in matters of artificial intelligence is the legal dimension of the processes in all their dimensions. In the middle of the twenty-first century, this remains one of the biggest question marks: "the legal issues involved with the development and implementation of artificial intelligence algorithms that we do not and cannot understand are substantial" (Price II, 2017, p. 6).

6 Final Considerations

The main objective of this chapter was to map digital transformation and artificial intelligence in higher education. The total of articles identified and adherent to the focus of this article were 13. To achieve this objective, a bibliometric analysis and a thematic analysis of the articles were outlined.

The chapter explored the application of artificial intelligence to higher education and detailed its transformative potential in the educational system. Through the use of advanced algorithms, machine learning, and natural language processing, artificial intelligence has the ability to personalize learning, improve administrative efficiency, and provide valuable insights to faculty and students.

It was found that the interest of researchers in the combination of transformation and artificial intelligence applied to higher education is still timid but growing. There are no author references in the Web of Science database.

Another advantage that is clear from this chapter is that the implementation of artificial intelligence in higher education can lead to a more adapted and inclusive educational experience, allowing students to develop their skills in a personalized way and at their own pace. There is a strategic and elastic dimension that allows adjusting syllabuses and pedagogical methodologies so that the teaching–learning process achieves the greatest effectiveness at each moment.

From another perspective, it was also concluded that the automation of routine administrative tasks allowed by the application of artificial intelligence to the more administrative and bureaucratic processes inherent to the operation of higher education can free up time and resources for more strategic and interactive activities, and thus improve the performance and overall effectiveness of the higher education institution.

In fact, automation of tasks is one of the main uses of artificial intelligence. For example, it can greatly increase the quality of research and investigation activities.

ChatGPT, for example, can support students and teachers in the process of identifying literature and scientific documents that focus on a particular subject under investigation. Tools like ChatGPT can even guide or even create summaries of articles and other academic and scientific documents. It can also help researchers to write drafts of various documents (grant applications, articles). When taken over by ChatGPT, these tasks free up the time of teachers, students, and researchers for other aspects of their work. Creativity and critical thinking are unique characteristics of human beings that will never be replaced by artificial intelligence. This also became quite clear in the course of writing this chapter.

Indeed, universities and all stakeholders directly or indirectly involved in processes of educational innovation should participate in the development and implementation of technology supporting artificial intelligence to better meet their needs.

There is no doubt that a culture favorable to the use of artificial intelligence in the university context can lead to the adoption of supplementary tools that contribute to the success of the teaching–learning process and all activities collateral to the area of education, namely, in areas such as inclusion or the production of "tailor-made" content. It is true that artificial intelligence, and tools like ChatGPT in particular, still have a long way to go towards full and undoubted affirmation. But it is also true that, today, there is no longer any doubt that artificial intelligence can definitely contribute to innovation in higher education and to increase the overall effectiveness of the teaching–learning process.

And, due to the unequivocal advantages associated with artificial intelligence, specifically applied to higher education, research, and scientific production in the area is already quite significant on a global scale. It was possible to confirm this in this chapter.

Regarding the "countries" that have more scientific production on the subject, Russia stands out with 20% of total publications. In second place, stands out Kingdom India with 17% of citations. However, there is an intense network of relationships established between the intensity of publication by country and the relationship established between them, through citations between works published in Brazil, the United Kingdom, Russia, Egypt, and the United Arab Emirates, placing them among the most productive countries.

In conclusion, to answer the research questions: (1) What is digital transformation?, (2) What is artificial intelligence? and (3) How can artificial intelligence technology be applied in higher education? Digital transformation in higher education is indicative of the need for a variety of solutions, including the use of technology (such as tablets and smartphones) in the learning process and cloud storage, virtual student assistance, and exam management through specific systems. The use of new technical resources is just one aspect of the digital revolution in higher education, which is a path without a clear end. Technology, education, and student needs need to be in harmony. Therefore, to achieve the best results, the HEI culture must be changed. The second question: What is artificial intelligence? AI is a field of study that focuses on artificially replicating the cognitive abilities of human intelligence to create software or machines capable of performing tasks typically performed by humans. The third question: How can artificial intelligence technology be applied in higher education? AI can be used in higher education for personalized education through the automation of administrative teaching tasks, software programs that favor the detection of topics that need classroom reinforcement, mentoring and supporting students outside of the classroom, and intelligently using data to teach and support students. In addition to personalized education, AI can be applied through augmented reality, adaptive platforms, intelligent tutoring systems, Chatbots, adaptive learning, Computer-Aided Instruction, direct feedback, Moocs, and Robotics. Two examples of success with the application of AI were at Georgia State University where AI was used to analyze large volumes of data, identify patterns, and provide personalized interventions, helping educational institutions to improve student performance and success. Another real case was at Stanford University, where a system called Stanford Automated Essay Scoring (SAES) was developed to assist in the automated correction of essays and essay evaluations.

However, it is important to note that artificial intelligence does not replace the teacher's importance in the educational process. The teacher's role becomes even more crucial, acting as a facilitator, guide, and motivator for the students.

Although there are clear benefits in applying artificial intelligence to higher education, a number of issues have also emerged clearly from this chapter that deserve reflection and further study. For example, ethical challenges such as data privacy and algorithm transparency need to be carefully addressed. It is crucial to ensure that the use of artificial intelligence in education is guided by sound ethical principles and that it promotes equity and inclusion.

Still, in short, artificial intelligence offers enormous potential to improve the quality and effectiveness of higher education by empowering both teachers and students. By strategically adopting these technologies, educational institutions can reap the benefits of a more personalized, efficient, and accessible education that prepares students for the challenges of today's world.

For future research, it is intended to map how the use of artificial intelligence and, for example, apps like ChatGPT can serve as a tool for sustainable technological innovation with the application of sustainable development objectives in the context of higher education.

References

Akour, M., & Alenezi, M. (2022). Higher education future in the era of digital transformation. *Education Sciences, 12*(11), 784.

Aldredge, M., Rogers, C., & Smith, J. (2021). The strategic transformation of accounting into a learned profession. *Industry and Higher Education, 35*(2), 83–88.

Alenezi, M. (2021). Deep dive into digital transformation in higher education institutions. *Education Sciences, 11*(12), 770.

Ali, M., Rahimi, R., Okumus, F., & Liu, J. (2016). Bibliometric studies in tourism. *Annals of Tourism Research, 61*, 180–198.

Baidoo-Anu, D., & Owusu, A. L. (2023). *Education in the era of generative artificial intelligence (AI): Understanding the potential benefits of ChatGPT in promoting teaching and learning*. Available at: https://ssrn.com/abstract=4337484

Bamiah, S. N., Brohi, S. N., & Rad, B. B. (2018). Big data technology in education: Advantages, implementations, and challenges. *Journal of Engineering Science and Technology, 13*(Special Issue on ICCSIT 2018), 229–241.

Baraishuk, D. (2021). *AI in higher education*. Belitsoft. https://belitsoft.com/custom-elearning-development/ai-in-education/ai-in-higher-education

Barzman, M., Gerphagnon, M., Aubin-Houzelstein, G., Baron, G. L., Bénart, A., Bouchet, F., et al. (2021). Exploring digital transformation in higher education and research via scenarios. *Journal of Futures Studies, 25*(3), 65–78.

Becker, S. A., Brown, M., Dahlstrom, E., Davis, A., DePaul, K., Diaz, V., & Pomerantz, J. (2018). *NMC horizon report: 2018 higher education edition*. Educause.

Berman, S. J. (2012). Digital transformation: Opportunities to create new business models. *Strategy & Leadership, 40*(2), 16–24.

Bonk, C. J., & Khoo, E. (2014). *Adding some TEC-VARIETY: 100+ activities for motivating and retaining learners online*. OpenWorldBooks.com and Amazon CreateSpace.

Borgman, C. L. (2017). *Big data, little data, no data: Scholarship in the networked world*. MIT press.

Bounfour, A. (2016). *Digital futures, digital transformation: From lean production to acceluction*. Springer International Publishing.

Bozanta, A., & Mardikyan, S. (2017). The effects of social media use on collaborative learning: A case of Turkey. *Turkish Online Journal of Distance Education, 18*(1), 96–110.

Bucea-Manea-Ţoniş, R., Martins, O. M. D., Bucea-Manea-Ţoniş, R., Gheorghiţă, C., Kuleto, V., Ilić, M. P., & Simion, V.-E. (2021). Blockchain technology enhances sustainable higher education. *Sustainability, 13*(22), 12347.

Bucea-Manea-Ţoniş, R., Kuleto, V., Gudei, S. C. D., Lianu, C., Lianu, C., Ilić, M. P., & Păun, D. (2022). Artificial intelligence potential in higher education institutions enhanced learning environment in Romania and Serbia. *Sustainability, 14*(10), 5842. https://doi.org/10.3390/su14105842

Daniel, B. (2015). B ig D ata and analytics in higher education: Opportunities and challenges. *British Journal of Educational Technology, 46*(5), 904–920.

de Bem Machado, A., Secinaro, S., Calandra, D., & Lanzalonga, F. (2022). Knowledge management and digital transformation for industry 4.0: A structured literature review. *Knowledge Management Research & Practice, 20*(2), 320–338.
Dodigovic, M. (2007). Artificial intelligence and second language learning: An efficient approach to error remediation. *Language Awareness, 16*(2), 99–113.
Dwivedi, Y. K., Hughes, D. L., Coombs, C., Constantiou, I., Duan, Y., Edwards, J. S., et al. (2020). Impact of COVID-19 pandemic on information management research and practice: Transforming education, work and life. *International Journal of Information Management, 55*, 102211.
Geerling, W., Mateer, G. D., Wooten, J., & Damodaran, N. (2023). ChatGPT has mastered the principles of economics: Now what? *SSRN Electronic Journal.*
Ghent, L., & Mateer, G. D. (2016). JET: A new journal takes off! *Journal of Economics Teaching, 1*(1), 0–6.
Gutiérrez-Salcedo, M., Martínez, M. Á., Moral-Munoz, J. A., Herrera-Viedma, E., & Cobo, M. J. (2018). Some bibliometric procedures for analyzing and evaluating research fields. *Applied Intelligence, 48*, 1275–1287.
Harvard Business Review. (2021). *How Georgia State University Increased Graduation Rates [Audio podcast].* https://hbr.org/podcast/2021/05/how-georgia-state-university-increased-graduation-rates
Hu, L. (2023). *Generative AI and Future.* Available at: https://pub.towardsai.net/generative-ai-and-future-c3b1695876f2
Kane, G. C., Palmer, D., Phillips, A. N., Kiron, D., & Buckley, N. (2015). Strategy, not technology, drives digital transformation. *MIT Sloan Management Review.*
Ke, Z., & Ng, V. (2019, August). Automated essay scoring: A survey of the state of the art. *IJCAI, 19*, 6300–6308.
Kuleto, V., Ilić, M., Dumangiu, M., Ranković, M., Martins, O. M., Păun, D., & Mihoreanu, L. (2021). Exploring opportunities and challenges of artificial intelligence and machine learning in higher education institutions. *Sustainability, 13*(18), 10424.
Kuleto, V., Bucea-Manea-Ţoniş, R., Bucea-Manea-Ţoniş, R., Ilić, M. P., Martins, O. M. D., Ranković, M., & Coelho, A. S. (2022). The potential of blockchain technology in higher education as perceived by students in Serbia, Romania, and Portugal. *Sustainability, 14*(2), 749.
Mandal, T., & Sarkar, S. (2020). An efficient block chain based security scheme for Internet of Vehicles (IoV): ICAISC-2020. In *International conference on recent trends in artificial intelligence, IOT, smart cities & applications*. SSRN Electronic Journal.
Marks, A., & AL-Ali, M. (2022). Digital transformation in higher education: A framework for maturity assessment. In M. Alaali (Ed.), *COVID-19 challenges to university information technology governance*. Springer.
Matt, C., Hess, T., & Benlian, A. (2015). Digital transformation strategies. *Business & Information Systems Engineering, 57*(5), 339–343.
Mills, T. (2018). AI vs AGI: What's the difference? *Forbes*. Available at: https://www.forbes.com/sites/forbestechcouncil/2018/09/17/ai-vs-agi-whats-thedifference/#517ec50d38ee
Mohamed Hashim, M. A., Tlemsani, I., & Matthews, R. (2021). Higher education strategy in digital transformation. *Education and Information Technologies*, 1–25.
Morakanyane, R., Grace, A. A., & O'Reilly, P. (2017). Conceptualizing digital transformation in business organizations: A systematic review of literature. In *Bled eConference* (pp. 427–444). Bled, Slovenia.
Moreno, V., Cavazotte, F., & Alves, I. (2017). Explaining university students' effective use of e-learning platforms. *British Journal of Educational Technology, 48*(4), 995–1009.
Ngabo, D., Wang, D., Iwendi, C., Anajemba, J. H., Ajao, L. A., & Biamba, C. (2021). Blockchain-based security mechanism for the medical data at fog computing architecture of internet of things. *Electronics, 10*(17), 2110.
Norvig, P., & Russell, S. (2013). Inteligência artificial. Rio de Janeiro: Grupo GEN.
Ouyang, F., Zheng, L., & Jiao, P. (2022). Artificial intelligence in online higher education: A systematic review of empirical research from 2011 to 2020. *Education and Information Technologies, 27*(6), 7893–7925.

Panchal, K., Shaikh, M., & Bilal, N. (2020). Artificial intelligence used in School's of China. In *Proceedings of the 3rd international conference on advance's in Science & Technology (ICAST) 2020*. Available at: https://doi.org/10.2139/ssrn.3560832,

Panchal, A. R., Bartos, J. A., Cabañas, J. G., Donnino, M. W., Drennan, I. R., Hirsch, K. G., ... & Berg, K. M. (2020). Part 3: adult basic and advanced life support: 2020 American Heart Association guidelines for cardiopulmonary resuscitation and emergency cardiovascular care. Circulation, 142(16_Suppl_2), S366-S468.

Petersen, J. (2021). *Innovative assessment practices*. Kindle Edition. Available at: https://learn.sd61.bc.ca/wp-content/uploads/sites/96/2017/09/FG-Innovative-Assessment-Whitepaper.pdf

Price, W. N., II. (2017). Artificial intelligence in healthcare: Applications and legal implications. *The SciTech Lawyer, 14*(1), 10–13.

Rodríguez-Abitia, G., & Bribiesca-Correa, G. (2021). Assessing digital transformation in universities. *Future Internet, 13*(2), 52.

Russell, S. J., & Norvig, P. (2010). Artificial intelligence a modern approach. London.

Santos, J. R. (2020). *Economia Indispensável*. Lisbon Press.

Selwyn, N. (2013). *Distrusting educational technology: Critical questions for changing times*. Routledge.

Sinha, B. (2020). Post-COVID challenges and opportunities in the education sector. *SSRN Electronic Journal*.

Smith, P., & Beretta, M. (2021). The gordian knot of practicing digital transformation: Coping with emergent paradoxes in ambidextrous organizing structures. *Journal of Product Innovation Management, 38*(1), 166–191.

Smutny, P., & Schreiberova, P. (2020). Chatbots for learning: A review of educational chatbots for the Facebook messenger. *Computers & Education, 151*, 103862.

Song, S., & Zhao, J. (2013). Automated essay scoring using machine learning. *International Journal of Advance Research, Ideas and Innovations in Technology, 5*(2), 1154–1156.

Surden, H. (2019). Artificial intelligence and law: An overview. *Georgia State University Law Review, 35*, 19–22.

Travis, D. B., Vail, M. W., & Anton, A. I. (2006, September). Towards regulatory compliance: Extracting rights and obligations to align requirements with regulations. In *14th IEEE International Requirements Engineering Conference (RE'06)* (pp. 49–58). IEEE.

Ting, D. S. W., Pasquale, L. R., Peng, L., Campbell, J. P., Lee, A. Y., Raman, R., ... & Wong, T. Y. (2018). Artificial intelligence and deep learning in ophthalmology. *British Journal of Ophthalmology*.

Verhoef, P. C., Broekhuizen, T., Bart, Y., Bhattacharya, A., Dong, J. Q., Fabian, N., & Haenlein, M. (2021). Digital transformation: A multidisciplinary reflection and research agenda. *Journal of Business Research, 122*, 889–901.

Wang, W., Huang, H., Zhang, L., & Su, C. (2021). Secure and efficient mutual authentication protocol for smart grid under blockchain. *Peer-to-Peer Networking and Applications, 14*, 2681–2693.

Westerman, G., Calméjane, C., Bonnet, D., Ferraris, P., & McAfee, A. (2011). Digital transformation: A roadmap for billion-dollar organizations. *MIT Center for Digital Business and Capgemini Consulting, 1*, 1–68.

Woods, R., Doherty, O., & Stephens, S. (2022). Technology driven change in the retail sector: Implications for higher education. *Industry and Higher Education, 36*(2), 128–137.

Zawacki-Richter, O., Marín, V. I., Bond, M., & Gouverneur, F. (2019). Systematic review of research on artificial intelligence applications in higher education – Where are the educators? *International Journal of Educational Technology in Higher Education, 16*(1).

Zeide, E. (2019). Artificial intelligence in higher education: Applications, promise and perils, and ethical questions. *Educause Review, 54*(3).

Zhong, J., Xie, H., Zou, D., & Chui, D. K. (2018, November). A blockchain model for word-learning systems. In *2018 5th international conference on Behavioral, Economic, and Socio-Cultural Computing (BESC)* (pp. 130–131). IEEE.

Education Institutions: Artificial Intelligence and Blockchain Higher

Andreia de Bem Machado, Leonor Domingos, Maria José Sousa, and Gertrudes Aparecida Dandolini

1 Context

The world of today is undergoing rapid change, and creativity is now a skill that is required across all industries and professions. In this context, it is clear that the multiple socio-ecological issues brought on by the digital era have created a variety of difficulties that call for novel solutions across all facets of society. This makes artificial intelligence (AI) an interdisciplinary field that focuses on the study and creation of tools and methods for simulating, enhancing, and extending human intelligence (Zhong et al., 2018). Blockchain technology can make relationships between parties in the virtual world safer, dependable, and with fewer intermediaries (Zhang et al., 2021a, b) in this technological scenario using AI. Due to these technologies, humanity is confronted with a new socio-technical paradigm that will have a profound impact on society at large and present prospects for transformation in many traditional economic areas. The teaching–learning process could benefit from innovation from both technologies (Zawacki-Richter et al., 2019).

Since its invention in 1956 by the Dartmouth Society, artificial intelligence technology has been used in an ever-increasing variety of spheres of human life, including education (Dodigovic, 2007). Building intelligent robots – or at least machines that act intelligently – is the goal of artificial intelligence. Artificial intelligence is a multidisciplinary field where technology is created to mimic human problem-solving abilities. Artificial neural networks, for instance, mimic a neuron's capacity

A. de Bem Machado (✉) · G. A. Dandolini
Federal University of Santa Catarina, Florianópolis, Brazil
e-mail: andreia.bem@ufsc.br; gertrudes.dandolini@ufsc.br

L. Domingos · M. J. Sousa
ISCTE – University Institute of Lisbon, Lisbon, Portugal
e-mail: leonor_domingos@iscte-iul.pt; maria.jose.sousa@iscte-iul.pt

A. de Bem Machado et al. (eds.), *Digital Transformation in Higher Education Institutions*, EAI/Springer Innovations in Communication and Computing,
https://doi.org/10.1007/978-3-031-52296-3_2

for learning (or a network of neurons). Artificial intelligence has entered an explosive stage in recent years as a result of the enormous strides achieved in machine learning, deep learning, and the exponential accumulation of data. Artificial intelligence has the potential to radically transform industries like security, banking, retail, transportation, and education because of its advantages in analysis, prediction, judgment, and decision-making. According to Fryer's et al., (2019) study article, which examined the function of chatbots in fostering students' interest in learning various languages, AI has specifically been employed in education both in the domain of educational management and in the teaching–learning process, notably in higher education.

Due to the development of trust for the recording and sharing of information in distributed systems, blockchain, a more modern technology, has also sparked interest in scientific talks for its use in higher education (Zhong et al., 2018). Other alluring advantages of this cutting-edge technology include decentralization, security, and data integrity.

As a result, these two technologies may work together because they both share the same needs for data analysis, security, and trust (Zhang et al., 2021a, b). For instance, the three essential components of artificial intelligence are algorithms, processing power, and data. The decentralized, immutable, and anonymous nature of blockchain, on the other hand, allows it to penetrate the data island and observe the movement of algorithms, computing power, and data resources. Additionally, blockchain can guarantee the original data's validity as well as the auditing of that validity and the traceability of artificial intelligence. Blockchain and artificial intelligence are considered cutting-edge technologies (Zhang, et al., 2021a, b) because of the vital roles they play in technical innovation and industrial transformation across a variety of situations, with the educational one standing out.

Therefore, the aim of this study is to map, through a scoping review, how blockchain and artificial intelligence technologies are being used or could be used in higher education. Six sections make up the research. The first is this introduction, followed by a discussion of blockchain concepts in the second, an explanation of artificial intelligence in the third, a discussion of the methodology used in this study in the fourth, a discussion of the study's findings in the fifth, and the authors' closing thoughts in the sixth.

Higher education institutions (HEIs) have been forced to redefine themselves and rethink in order to keep up with the technological challenges of the post-modern world in this scenario marked by the COVID-19 pandemic, linked to profound social, economic, and cultural transformations, and the galloping development of digital information and communication technologies. Reconsider innovative educational methods and procedures that use technological resources in classroom instruction. In light of this, the chapter's central issue was how to use AI and blockchain technology in higher education. This will be addressed with the following goal: to outline how Blockchain and Artificial Intelligence can be used in education as technology for higher education through a scoping review. As a result, in addition to this introduction section, the article is divided into five further sections.

2 Blockchain

Blockchain technology must be conceptualized in terms of a system that enables the decentralized creation of distributed records of digital events. Information can be shared transparently thanks to this sophisticated database method (Alamary et al., 2019). As a result, decentralization, security and integrity, anonymity, and encryption are all associated with blockchain technology. It can also be viewed as a consensus mechanism that rewards universities, academic staff, and students through a smart contract. Therefore, according to Di Pierro (2017), this technology has helped to improve higher education.

Researchers, technologists, and businesspeople have recently focused their attention on the blockchain topic in this scenario. The development of trust for information recording and sharing in distributed systems is the key driver of interest (Di Pierro, 2017). Other alluring advantages of cutting-edge technology include decentralization, security, and data integrity. Although technology has opened up new opportunities for several facets of society, little research has been done on blockchain practices in relation to educational endeavors. Preliminary scientific investigations in this context, taking into account the analytical goals of this research, focus on issues like the development of applications, analysis of the advantages of blockchain for educational institutions and other parties involved, as well as highlighting the difficulties of using blockchain in education.

Studies, like the one by Alamary et al. (2019), give unified findings on numerous blockchain technology uses in the context of education. In conclusion, the following groups stand out: managing certificates, managing competencies and learning outcomes, evaluating students' professional aptitude, protecting the collaborative learning environment, protecting learning objects, transferring fees and credits, getting digital guardianship consent, managing competitions, managing copyrights, enhancing student interaction on e-learning, reviewing exams, and supporting students' lifelong learning are all among the tasks.

Researchers are excited about the potential of developing concepts for the construction of "mega-universities" as one of the blockchain application options. According to Skiba (2017), educational institutions can build ecosystems using blockchain technology that enables students to design their own learning routes in conjunction with the responsible professors in a cooperative exchange of experiences. It is clear that this is a difficult journey that requires a number of steps, not only in terms of education but also in terms of political acceptance and consensus among international institutions.

Yumna et al. (2019) provide solutions to problems concerning educational institutions and the usage of resources compatible with blockchain technology in this scenario for education. The severity of physical, digital, and financial issues in educational institutions is taken into account by these authors. Their research suggests that the primary obstacle with blockchain applications has to do with the issue of record exchange and verification across organizations. In contrast, the report highlights the advantages of the technology, such as decentralization, traceability, and a consensus mechanism for the veracity of data and information – all of which are inherent characteristics of blockchain. The authors also provide a warning regarding a few technical issues that are thought to be obstacles to the use of technology in the educational setting, including (a) the immutability of records, (b) the disclosure of personal information, and (c) the system's scalability.

The use of blockchain technology in education is showing encouraging results recently, with an emphasis on current uses and future prospects, showcasing a potential that can greatly help the education sector. According to Bhaskar (2021), the use of blockchain technology in education includes several qualitative aspects. Additionally, according to Bhaskar (2021), Alammary et al. (2019), and Sun et al. (2018), among the advantages of blockchain technology in the education sector, the following are the most obvious:

(1) Managing and verifying data while maintaining authenticity.
(2) Blockchain data is readily available and transparently verifiable around the clock.
(3) Certificates of education are frequently issued and verified using the blockchain. Degrees, transcripts, student competences, accomplishments, and professional aptitude, information that companies worldwide may verify.
(4) The certification procedure is streamlined by blockchain technology.
(5) Less time must be spent by the employer in confirming academic findings.
(6) A safe platform for sharing student data is provided by blockchain, boosting confidence, lowering costs, and promoting transparency.
(7) Using data blocks and timestamps, blockchain technology maintains a complete record of the course.
(8) Old and fresh data blocks in the blockchain cannot be erased because of the records' immutability.
(9) The cryptographic technique decreases the likelihood of fraud and prevents data manipulation.
(10) Blockchain develops a virtual infrastructure for managing credentials, storing documents, and keeping track of student accomplishments.

According to the research of the authors Sun et al. (2018), Alammary et al. (2019), Braskar et al. (2021), and Sun et al. (2018), these benefits enable public policymakers and other stakeholders as well as providing subsidies for educational institutions. Therefore, by utilizing blockchain technology, it is possible to construct an innovation ecosystem for applications in educational systems, igniting new horizons for development in education-related fields, and pursuing a sustainable future for systems of social inclusion and education.

Based on this data, it is crucial to think seriously about conducting a full examination of the problems related to the adoption of that technology. That is, to create strong foundations through research that comes before building, maintaining, and managing educational platforms on blockchain. By doing this, the range of applications could expand beyond the ones that are already used, such as maintaining records, creating and managing certificates, and creating decentralized educational ecosystems.

3 Artificial Intelligence

John McCarthy was the first to define artificial intelligence in a two-month workshop at Dartmouth College in the US, despite the fact that several definitions have surfaced in recent years. Artificial intelligence (AI) is a branch of study that focuses on artificially imitating human intelligence's cognitive capacities to build computer programs or other devices that can carry out tasks that are traditionally handled by people. They explicitly state that AI is the science and engineering of creating intelligent machines, particularly intelligent computer programs, in their definition of the term. While it is related to the related job of utilizing computers to study human intelligence, AI need not be constrained to techniques that can be observed biologically.

AI will function under the presumption that, in theory, every facet of learning or any other characteristic of intelligence can be sufficiently correctly characterized for a machine to replicate. There will be an effort to figure out how to make machines understand language, create abstractions and notions, solve human-only problems, and advance (Zawacki-Richter et al., 2019).

AI is defined by Dodigovic (2007) as machines that carry out cognitive processes that are usually similar to those of the human mind, especially learning and problem-solving. Additionally, they clarify that AI does not refer to a specific technology. It serves as a catch-all phrase for a variety of technologies and approaches, including genetic algorithms, machine learning, natural language processing, data mining, and neural networks.

A different way to define AI is as a group of models, techniques, and technologies (such as knowledge searching, reasoning, and representation, decision mechanisms, perception, planning, natural language processing, uncertainty handling, and machine learning) that, either individually or collectively, address issues requiring logical reasoning (Woods et al., 2022).

As a result, the term "artificial intelligence" (AI) is wide and covers a variety of concepts, including "knowing," "reasoning," "problem solving," "perception," "learning," "planning," and the capacity to control and move objects" (Woods et al., 2022). The presence of AI in various situations, including the educational context – which has progressively integrated technology instruments into the pedagogical process – is taken into account in this scenario.

4 Methodology

The systematic scoping review methodology was used in this study to consider many data sources and types of publications, both theoretical and empirical, in order to better comprehend a concept or phenomenon (Tricco et al., 2018). For scoping reviews, Tricco et al. (2018) modified the PRiSMA approach and called it PRISMA-ScR. A protocol for conducting scoping reviews ensures higher methodological rigor for the research and improves its applicability, transparency, and repeatability.

The search used descriptors that allowed it to cover the key research terms and their corresponding synonyms used in the literature. Because they are multidisciplinary and use a peer review process, two databases (Scopus and Web of Science) were selected. The selection of these bases for electronic searches is justified by the fact that Web of Science (Clarivate Analytics) and Scopus (Elsevier) are currently the largest indexing services, with the majority of publications focusing on artificial intelligence and higher education (Mongeon & Paul-Hus, 2016; Ouyang et al., 2022).

Following the flowchart of PRISMA-ScR, it was sought to fulfill the steps of identification, selection, eligibility, and inclusion of articles, as shown in Table 1.

The methodological steps of the PRISMA-ScR protocol involve the determination and justification of the eligibility criteria, the description of the information sources, the presentation of the database search strategy, the data presentation process, and the synthesis of the results. All steps are presented below.

4.1 Steps

4.1.1 Step 1: Identification

In the identification step, the search terms used were "artificial intelligence," "blockchain," and "higher education." The searches were conducted between 06/15/2022 and 07/04/2022. A total of 71 articles were identified, distributed into 38 articles from Web of Science and 33 from Scopus. Then only articles from peer-reviewed academic journals were included to ensure scientific rigor, which resulted in 71 articles.

Table 1 Prisma scoping review

Prisma scoping review					
Search terms ("artificial intelligence" and "blockchain" and "higher education")					
Research carried out in the SCOPUS and WOS databases between June and July 2022					
Identification	Total found in the databases		SCOPUS	WoS	Total
			33	38	71
Selection	Inclusion criteria	Publication type: Peer-reviewed review and research articles	8	10	18
		Connection with the terms AI, blockchain, and higher education.	8	10	18
	Exclusion criteria	Publication type: Books and articles from event proceedings	25	28	53
Eligibility	Analysis of titles and abstracts and cross-referencing between databases to remove duplicates		8	10	18
Inclusion	Total included		3	10	13

Source: Prepared by the authors (2022)

4.1.2 Step 2: Selection

In order to obtain the largest number of publications, only two inclusion criteria were adopted: (1) publications must be available as "open access", in line with the growing worldwide movement of open access policies for research; (2) research must be classified as peer-reviewed articles, due to the nature of this article: in systematic reviews, it is recommended to analyze only this type of document. In the selection step, articles were first selected by reading the titles, abstracts, and keywords, leaving 38 publications in Scopus and 33 in Web of Science for the next step. Of these, five articles appeared duplicated in both databases, resulting in 66 articles selected.

4.1.3 Step 3: Eligibility

As eligibility criteria, in this step, the titles and abstracts were analyzed and the databases were cross-referenced to remove duplicates. After the screening, eight selected articles followed for full-text analysis that met the eligibility criteria in the Scopus database and 10 in the Web of Science database. Since five articles are duplicates in both databases, a total of 13 articles resulted for full-text reading.

4.1.4 Step 4: Inclusion

Finally, 13 articles were included in the qualitative and quantitative syntheses. In the PRISMA recommendation, there is no rule regarding the minimum or maximum number of articles to be included. According to Moher et al. (2009), the review team must search the available literature. This search results in the number of reports found. Once these reports were screened and the eligibility criteria applied, a smaller number of articles remained. As research inclusion criteria, the following were defined: (a) peer-reviewed articles, in order to ensure the quality of the selected publications; (b) the descriptors should appear in the title, abstract or keywords; (c) without chronological delimitation; (d) with a connection to the researched theme (blockchain and artificial intelligence in higher education). The exclusion criteria were: (a) books and conference papers; (b) articles without adherence to the research objectives; (c) unavailability of full access to the article, either by the base itself, contact with authors, or through parallel platforms such as Google Scholar, Research Gate, and Emerald Insight.

Thus, first, a bibliometric analysis was performed (Sect. 4.1), and then a careful reading of the selected publications in order to answer the research question. From the thematic analysis, three categories of analysis were defined: AI – applicability in Higher Education, Blockchain – applicability in Higher Education, and AI and Blockchain – applicability in Higher Education, which are described in Sect. 4.2.

4.2 Analyses of the Articles

The 13 publications listed in Table 1 were read and examined in order to respond to the research question, "How might blockchain technology and artificial intelligence be applied to higher education?" (Table 2)

Eight of the 13 publications that made up the corpus of analysis were comprehensive reviews of the literature, and five of them dealt with empirical investigations utilizing questionnaires. In light of the primary study question ("How might Blockchain technology and artificial intelligence be applied to higher education?"), which will be examined and answered below, it was determined that specific topics jumped out after reading and analyzing the papers.

4.2.1 AI and Blockchain – Applicability to Higher Education

To provide equal and inclusive access to education, blockchain and AI technology can be applied in higher education (Reis-Marques et al., 2021). They can aid in cost savings and the creation of a sustainable, all-inclusive educational system. By installing blockchain applications in a variety of domains (social media, education, cryptocurrencies, technology, and data management), including building intelligent learning environments with the aid of AI technology and developing educational software products, blockchain can ensure information security and document verification processes in higher education institutions, ensuring inclusion and equitable access to education (Reis-Marques et al., 2021).

The implementation of Adaptive Platforms can use both AI and blockchain (Reis-Marques et al., 2021). These are tools that allow students to follow their own unique knowledge trails, beginning with previously learned material, in an effort to foster their cognitive growth and independence. Universities should build a clear power structure and effective artificial intelligence-based communication platforms in the context of higher education (Wang et al., 2020). In addition, it is feasible to improve the format and duration of information and include learning tools by employing MOOCs, augmented reality, virtual reality, gamification, blockchain, and video conferencing (Woods et al., 2022).

In this sense, technological disruptions include developments in the fields of automation and machine learning, artificial intelligence, data analytics, and blockchain (Aldredge et al., 2021). In addition to the AI story, applying blockchain to higher education results in the fusion of the most diversified populations, cultures, and educational levels (Kataev & Bulysheva, 2022). Since learning at higher education institutions can be favored by new technologies, blockchain and AI can be linked to the idea of sustainability in this way (Bucea-Manea-oniş et al., 2022). Blockchain will make sure that data in the educational system is dependable, safe, and private. On the other side, AI models will use this data to expand the options for teaching–learning approaches in institutions of higher learning (Chivu et al., 2022).

Another way that AI and blockchain might be used in higher education is through Direct Feedback, which has been linked to increased involvement, improved performance, and better engagement among teachers and students since it can give

Table 2 Articles for scoping review

ID	Year	Authors	Article
1	2019	Williams, P.	Does competency-based education with blockchain signal a new mission for universities?
2	2020	Aldredge M; Rogers C; Smith J	The strategic transformation of accounting into a learned profession
3	2020	Hernandez-de-Menendez M; Escobar Daz C; Morales-Menendez R	Technologies for the future of learning: State of the art
4	2021	Reis-Marques C; Figueiredo R; De Castro Neto M	Applications of Blockchain technology to higher education arena: A bibliometric analysis
5	2021	Ayub Khan A; Laghari AA; Shaikh S; Mamlouk AM; Alshazly H	Educational Blockchain: A secure degree attestation and verification traceability architecture for higher education commission
6	2021	Woods R; Doherty O; Stephens S	Technology driven change in the retail sector: Implications for higher education
7	2021	Ngabo D; Wang D; Iwendi C; Anajemba JH; Ajao LA; Biamba C	Blockchain-based security mechanism for the medical data at fog computing architecture of internet of things
8	2021	Bucea-Manea-Tonis R; Martins R; Gheorghita C; Kuleto MP; Simion V.E.	Blockchain technology enhances Sustainable higher education
9	2022	Al-Ghamdi Asam; Ragab M	Artificial Intelligence techniques based learner authentication in cybersecurity higher education institutions
10	2022	Bucea-Manea-Tonis R; Kuleto V; Gudei SCD; Lianu C; Lianu C; Ilic MP; Paun D	Artificial Intelligence potential in higher education Institutions enhanced learning environment in Romania and Serbia
11	2022	Kuleto V; Bucea-Manea-Tonis R; Bucea-Manea-Tonis R; Ilic MP; Martins OMD; Rankovic M; Coelho AS	The potential of Blockchain Technology in Higher Education as perceived by students in Serbia, Romania, and Portugal
12	2022	Kataev M; Bulysheva L	Blockchain system in the higher education: Storing academical students' records and achievements accumulated in the educational process
13	2022	Chivu RG; Popa IC; Orzan MC; Marinescu C; Florescu MS; Orzan AO	The role of Blockchain Technologies in the Sustainable Development of students' learning process

Source: Prepared by the authors (2022)

feedback based on errors committed during an activity. Students can complete the activities and exercises by receiving feedback that evaluates the incorrect questions (Kuleto et al., 2022). Giving pupils quick feedback on their learning progress is one of AI‘s most significant contributions to education and meaningful learning. Through immediate feedback that encourages students to become more involved, active learners, and autonomous, it may thus encourage more student engagement, increase performance, motivation, and self-regulation. Blockchain in higher education has shown that student motivation has a large and favorable impact on the caliber of collaborative student work. It has been demonstrated that group work increases student involvement in the learning process, and improved learning outcomes for the students involved (Hernandez-de-Menendez et al., 2020).

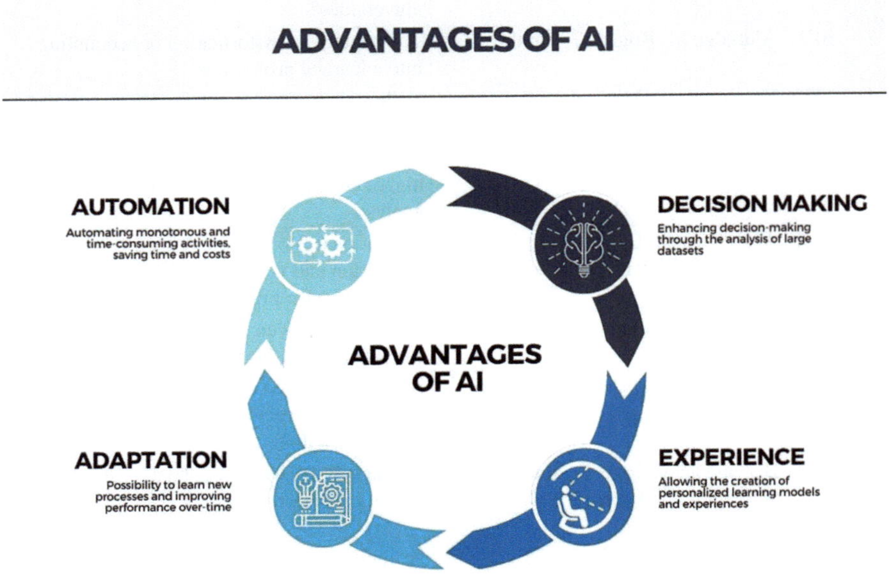

Figure SEQ Figure*ARABIC 1 – Representation of the main advantages of AI (Source: Prepared by the authors, [2023])

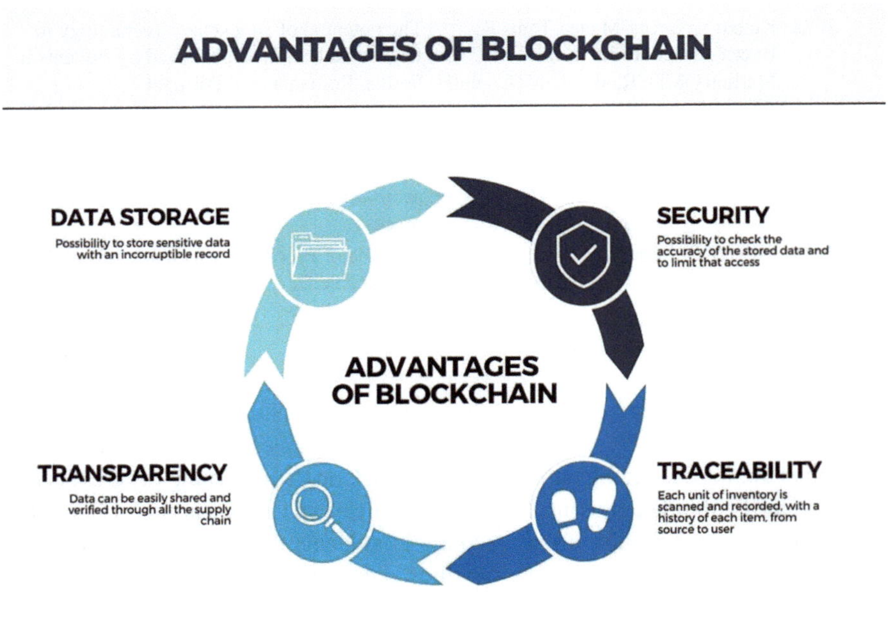

Figure SEQ Figure*ARABIC 2 – Representation of the main advantages of Blockchain (Source: prepared by the authors, [2023])

5 Final Conclusions

This article's major goal was to use a scoping review based on the PRISMA-ScR technique to map in the Scopus and Web of Science databases how blockchain and artificial intelligence can be used in education as a technology for higher education. There were a total of 13 articles found that matched the theme of this page. A bibliometric analysis and a thematic analysis of the papers were performed in order to accomplish this objective.

It was discovered that scholars' interest in the application of blockchain and artificial intelligence to higher education is still hesitant but expanding. Radu Bucea-Manea-Oniş, a reference author with five citations in Scopus and four in WoS, is one of the authors who received the most citations.

Regarding the nations, Romania and Russia had the most scientific output on the topics of blockchain and higher education, each with three papers in Scopus. However, due to the strong relationship network that has been built between the nations as a result of citations between works published in the Scopus and Web of Science, several other European, North American, and African nations have indicated interest in the issue.

As a result, three themes – the applicability of artificial intelligence to higher education, the applicability of blockchain to higher education, and the joint application of AI and blockchain to higher education – were discovered after conducting a thematic analysis of the articles in order to respond to the research question about how artificial intelligence and blockchain technologies can be applied to higher education. Several objects and functions, including adaptive platforms, intelligent tutoring systems, engagement, interactive chatbots, adaptive learning, computer-aided instruction, direct feedback, MOOCs, and robotics, were discovered through this analysis.

The relationship between blockchain and artificial intelligence was also discovered to be important and complimentary, which adds additional value to the innovation process in higher education and takes into account potential applications for the needs of students. In the same way, blockchain technology is required in the educational sector since it plays a critical role in the security process, particularly in authenticating documents that include academic facts and in providing a dependable solution to stop any academic fraud. Future studies will map out how blockchain technology and artificial intelligence might be used to advance sustainable technological innovation and advance sustainable development objectives in the context of higher education.

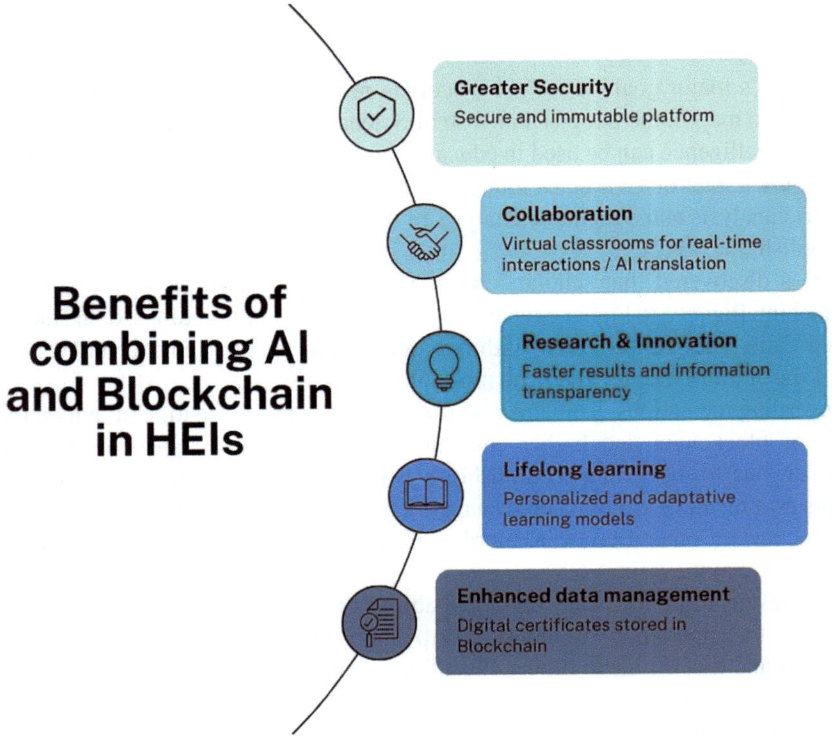

Figure SEQ Figure*ARABIC 3 – Benefits of combining AI and Blockchain technologies in HEIs (Source: prepared by the authors, [2023])

6 Future Possibilities

The unprecedented use of AI and blockchain in the field of higher education is allowing HEIs to change the paradigm of education, and to change how learning, administration, and accreditation are being done at the moment. The traditional models of education and learning can be replaced by more dynamic and personalized learning experiences. AI algorithms, powered by vast datasets and advanced analytics, have changed how content is delivered and assessed to and by people. It's possible, that in the future students will have an AI-powered tutor, which can adapt to each student's learning process, pace, weakness, and strengths, tailoring classes' content and adapting assignments for maximum comprehension and engagement. AI also allows for real-time feedback and adaptative learning, this way all students can learn in their own personalized way, whether they have difficulties or they aim for harder challenges. On the other hand, blockchain can change the way higher education administration works, by making the process of verifying people's credentials and certificates easier and automated. Since blockchain technology provides a secure and immutable platform for recording and sharing academic achievements, HEIs can issue digital certificates that are stored in blockchain, providing instant tamper-proof access to anyone that needs to verify a student's credentials. This will result in reduced administrative labor, but also ensure the authenticity and integrity of academic records, mitigating the possibilities of fraudulent degrees.

These two technologies have also aided in enabling cross-country collaboration thus offering further access to education, for those who could not access it. Now, it's possible to create virtual classrooms, powered by AI-driven language translation and immersive technologies, which allow for real-time interactions between students and their teachers, wherever they are physically. Blockchain-enabled micropayments and smart contracts can also make education more affordable and accessible, as students can be given the possibility of paying for individual modules or courses, instead of a full-time course. This makes it possible to democratize knowledge and allow people to be lifelong learners.

The synergy between AI and Blockchain can also develop research and innovation within HEIs, as AI- algorithms can analyze vast amounts of data, accelerating the time required for research, thus achieving results faster. Blockchain technology ensures the transparency, traceability, and reproducibility of research findings, contributing to putting a stop to data manipulation and academic misconduct. Having collaborative and global research networks, supported by smart contracts, facilitates the sharing of resources and knowledge, allowing for greater open innovation.

The convergence of AI and Blockchain allows for a paradigm- shift from the traditional degree-centric education to a lifelong continuous learning and skill adaptation. AI platforms can greatly aid students to assess their skills, identify knowledge gaps, and have access to personalized learning models. Blockchain-powered credentials are updated in real time, reflecting the latest skills and competencies acquired by an individual.

Lastly, the future of higher education and HEIs is also dependent on the nexus of AI and Blockchain. Both technologies have the potential to democratize education, enhance transparency, accelerate research, and foster global communities of learners and educators. While challenges and ethical concerns persist, there's also the possibility of a more accessible, personalized, and adaptable education system.

References

Alammary, A., Alhazmi, S., Almasri, M., & Gillani, S. (2019). Blockchain-based applications in education: A systematic review. *Applied Sciences (Basel, Switzerland), 9*(12), 2400. https://doi.org/10.3390/app9122400

Aldredge, M., Rogers, C., & Smith, J. (2021). The strategic transformation of accounting into a learned profession. *Industry and Higher Education, 35*(2), 83–88. https://doi.org/10.1177/0950422220954319

Bhaskar, P., Tiwari, C. K., & Joshi, A. (2021). Blockchain in education management: Present and future applications. *Interactive Technology and Smart Education, 18*(1), 1–17.

Bucea-Manea-Ţoniş, R., Kuleto, V., Gudei, S. C. D., Lianu, C., Lianu, C., Ilić, M. P., & Păun, D. (2022). Artificial intelligence potential in higher education institutions enhanced learning environment in Romania and Serbia. *Sustainability, 14*(10), 5842. https://doi.org/10.3390/su14105842

Chivu, R.-G., Popa, I.-C., Orzan, M.-C., Marinescu, C., Florescu, M. S., & Orzan, A.-O. (2022). The role of blockchain technologies in the sustainable development of students' learning process. *Sustainability, 14*(3), 1406. https://doi.org/10.3390/su14031406

Di Pierro, M. (2017). What is the blockchain? *Computing in Science & Engineering, 19*(5), 92–95.

Dodigovic, M. (2007). Artificial intelligence and second language learning: An efficient approach to error remediation. *Language Awareness, 16*(2), 99–113. https://doi.org/10.2167/la416.0

Fryer, L. K., Nakao, K., & Thompson, A. (2019). Chatbot learning partners: Connecting learning experiences, interests and competence. *Computers in Human Behavior, 93*, 279–289.

He, Y., Wang, Y., Qiu, C., Lin, Q., Li, J., & Ming, Z. (2021). Blockchain-based edge computing resource allocation in IoT: A deep reinforcement learning approach. IEEE internet of things journal, 8(4), 2226–2237. https://doi.org/10.1007/10.1109/jiot.2020.3035437

Hernandez-de-Menendez, M., Escobar Díaz, C., & Morales-Menendez, R. (2020). Technologies for the future of learning: State of the art. *International Journal on Interactive Design and Manufacturing (IJIDeM), 14*(2), 683–695. https://doi.org/10.1007/s12008-019-00640-0

Kataev, M., & Bulysheva, L. (2022). Blockchain system in the higher education: Storing academical students' records and achievements accumulated in the educational process. *Systems Research and Behavioral Science, 39*(3), 589–596. https://doi.org/10.1002/sres.2872

Kuleto, V., Bucea-Manea-Ţoniş, R., Bucea-Manea-Ţoniş, R., Ilić, M. P., Martins, O. M. D., Ranković, M., & Coelho, A. S. (2022). The potential of blockchain technology in higher education as perceived by students in Serbia, Romania, and Portugal. *Sustainability, 14*(2), 749. https://doi.org/10.3390/su14020749

Moher, D., Liberati, A., Tetzlaff, J., Altman, D. G., & PRISMA Group*. (2009). Preferred reporting items for systematic reviews and meta-analyses: the PRISMA statement. *Annals of Internal Medicine, 151*(4), 264–269.

Mongeon, P., & Paul-Hus, A. (2016). The journal coverage of Web of Science and Scopus: a comparative analysis. *Scientometrics, 106*, 213–228.

Ouyang, F., Zheng, L., & Jiao, P. (2022). Artificial intelligence in online higher education: A systematic review of empirical research from 2011 to 2020. *Education and Information Technologies, 27*(6), 7893–7925.

Reis-Marques, C., Figueiredo, R., & de Castro Neto, M. (2021). Applications of Blockchain technology to higher education arena: a bibliometric analysis. *European Journal of Investigation in Health, Psychology and Education, 11*(4), 1406–1421.

Skiba, D. J. (2017). The potential of blockchain in education and health care. *Nursing Education Perspectives, 38*(4), 220–221. https://doi.org/10.1097/01.NEP.0000000000000190

Sun, H., Wang, X., & Wang, X. (2018). Application of blockchain technology in online education. *International Journal of Emerging Technologies in Learning (IJET), 13*(10), 252. https://doi.org/10.3991/ijet.v13i10.9455

Tricco, A. C., et al. (2018). PRISMA extension for scoping reviews (PRISMA-ScR): Checklist and explanation. *Annals of Internal Medicine, 169*(7), 467–473. https://doi.org/10.7326/M18-0850. Disponível em: Disponível em: https://www.acpjournals.org/doi/10.7326/M18-0850. Acesso em: 16 jun. 2021

Wang, W., Huang, H., Zhang, L., & Su, C. (2020). Secure and efficient mutual authentication protocol for smart grid under blockchain. *Peer-to-Peer Networking and Applications, 14*, 2681. https://doi.org/10.1007/s12083-020-01020-2

Woods, R., Doherty, O., & Stephens, S. (2022). Technology driven change in the retail sector: Implications for higher education. *Industry and Higher Education, 36*(2), 128–137. https://doi.org/10.1177/09504222211009180

Yumna, H., et al. (2019). Use of blockchain in education: A systematic literature review. In *Asian conference on intelligent information and database systems* (pp. 191–202). Springer.

Zawacki-Richter, O., Marín, V. I., Bond, M., & Gouverneur, F. (2019). Systematic review of research on artificial intelligence applications in higher education – Where are the educators? *International Journal of Educational Technology in Higher Education, 16*(1). https://doi.org/10.1186/s41239-019-0171-0

Zhang, L., Peng, M., Wang, W., Jin, Z., Su, Y., & Chen, H. (2021a). Secure and efficient data storage and sharing scheme for blockchain-based mobile-edge computing. *Transactions on Emerging Telecommunications Technologies, 32*(10). https://doi.org/10.1002/ett.4315

Zhang, L., Zou, Y., Wang, W., Jin, Z., Su, Y., & Chen, H. (2021b). Resource allocation and trust computing for blockchain-enabled edge computing system. *Computers & Security, 105*(102249), 102249. https://doi.org/10.1016/j.cose.2021.102249

Zhong, J., et al. (2018). A blockchain model for word-learning systems. In *2018 5th International Conference on Behavioral, Economic, and Socio-Cultural Computing (BESC)* (pp. 130–131). IEEE.

Digital Transformation in Higher Education Institutions: A Case Study at Polytechnic University of Tomar

Célio Gonçalo Marques, Lígia Mateus, and Inês Araújo

1 Introduction

Digital transformation plays a crucial role in optimizing the quality of teaching and preparing students for an increasingly technology-driven world (Nóvoa et al., 2023). As stated in the Valencia 2023 Declaration: "We live in an interconnected and complex world where technology is advancing exponentially. We face enormous social and environmental challenges that urgently require creative, innovative, and transformative solutions" (Declaración de Valencia: Universidad y Sociedad, 2023). Technology allows educational institutions to transcend physical limits and connect students and teachers from different parts of the world (Hashim et al., 2022). Not only does it make learning more accessible to a wider audience, but it also allows for personalization of teaching, adapting content and teaching strategies to the individual needs of students and providing interactive and engaging learning experiences (Fernández et al., 2023).

On a global scale, digital transformation has the potential to break down geographical boundaries, enabling organizations, companies, and individuals to connect and collaborate on a global level, facilitating knowledge sharing, collaboration on international projects, and the creation of global networks (Safonov et al., 2022; Ydyrysbayev et al., 2022). Digital transformation also increases the efficiency of global operations (Feliciano-Cestero et al., 2023; Veldhoven & Vanthienen, 2022). Automated processes, advanced data analytics, and new technologies improve productivity and real-time decision-making – essential factors for competing in a dynamic global scenario (Kraus et al., 2021). On the business front, digital

C. G. Marques (✉) · L. Mateus · I. Araújo
Laboratory of Pedagogical Innovation and Distance Learning (LIED), Polytechnic University of Tomar, Tomar, Portugal
e-mail: celiomarques@ipt.pt; ligia.mateus@ipt.pt; inesaraujo.lied@ipt.pt

A. de Bem Machado et al. (eds.), *Digital Transformation in Higher Education Institutions*, EAI/Springer Innovations in Communication and Computing,
https://doi.org/10.1007/978-3-031-52296-3_3

transformation is a catalyst for growth (Ellström et al., 2022; Nadkarni & Prügl, 2021). It opens doors to new business models, fosters innovation, and increases competitiveness in the global marketplace (Konopik et al., 2022). On the social front, it can promote inclusion and equality by providing access to educational resources, health services, and employment opportunities around the world (Mukul & Büyüközkan, 2023). The importance of digital transformation from a global perspective lies in its ability to connect, empower, and transform the way societies, organizations, and individuals operate on a global scale (Marks et al., 2020).

The availability of a wide range of online educational resources enriches the content of teaching and expands learning opportunities. In addition, technology facilitates and encourages collaboration between different higher education institutions, promoting the exchange of best practices, perspectives, and knowledge between professionals and students (Mukul & Büyüközkan, 2023).

We have now succeeded in creating digital repositories that store and deliver information efficiently, facilitating access to academic, scientific, and cultural resources. These repositories transcend geographical boundaries, enabling the global dissemination of knowledge that is essential for rapid advances in research and development. UNESCO's "Global Education Monitoring Report, 2023: Technology in Education" (GEM Report 2023) presents some interesting data on this, such as "90% of content in higher education repositories with open education resource collections was created in Europe and Northern America and 92% of content in the OER Commons global library is in English" (*Global Education Monitoring Report, 2023: Technology in Education: A Tool on Whose Terms?*, 2023). The creation of Massive Open Online Courses (MOOCs) democratizes access to education by allowing people in different parts of the world to participate in courses offered by renowned institutions, removing physical and economic barriers. According to the same document, higher education is the fastest adopter of digital technology and the most transformed by it. In 2021, there were over 220 million students attending MOOCs. These digital platforms offer personalized learning that adapts to each student's pace and style, making education more flexible and adaptable to individual needs. Digital transformation is revolutionizing the way we store, share, and access information, fostering global collaboration, and making education more accessible and personalized. These changes are key to driving innovation, growth, and equality in today's society.

As well as the benefits of digital transformation in education, there are also significant challenges (Alenezi, 2021; Marks et al., 2020). Inequality of access, whether due to a lack of digital devices or Internet connectivity, creates a 'digital divide' between students (McCarthy et al., 2023), resulting in some young people having more digital learning opportunities than others. The aforementioned GEM Report 2023 highlights the inequalities caused by digital learning, especially during the COVID-19 pandemic, where 500 million students, mainly the poorest and those in rural areas, have been neglected due to the transition to digital education (Mhlanga & Moloi, 2020). A significant imbalance in online resources has been observed, favoring Europe and North America (*Global Education Monitoring Report, 2023: Technology in Education: A Tool on Whose Terms?*, 2023).

Promoting digital literacy in higher education plays a key role in preparing students for today's labor market, where digital skills are increasingly valued (Bucăţa et al., 2022; Safonov et al., 2022). In its "Skills Outlook 2023" report, the Organisation for Economic Co-operation and Development (OECD) highlights the importance of investing in skills "to build a resilient green and digital transition," harnessing the potential of technology and artificial intelligence. The organization points out that only around 40% of adults participate in work-related training, making it difficult to upgrade and reskill. It stresses the need to align education and training with labor market needs to ensure that everyone can participate in and benefit from economic development, especially in the face of environmental and digital change. The OECD stresses that education systems must step up efforts to develop the skills of young people, as only one in three combines scientific literacy with the attitudes needed to be conscious consumers and workers in the green economy. In short, the OECD advocates empowering the whole population, including young people in initial education and adults in lifelong learning (OECD Skills Outlook 2023, 2023).

Digital transformation not only improves the quality of teaching but also empowers students to face the challenges of a constantly evolving world (*European Commission. Digital Education Action Plan*, 2018; McCarthy et al., 2023). Ensuring that digital educational resources are accurate, up-to-date, and effective is a critical element in ensuring the quality of learning (Quaicoe et al., 2023). It is also essential to develop effective assessment methods that are adapted to the digital environment.

For its part, the GEM Report 2023 from UNESCO highlights the intense debate about the role of technology in education, essentially providing an in-depth analysis of the role of technology in education and highlighting concerns about the excessive use of technology in schools. The document recommends a "human-centered vision" in the use of technology in education and highlights the gaps in access. According to UNESCO, information and communication technologies have the potential to promote equity and inclusion by reaching disadvantaged students and disseminating knowledge in accessible formats. In some cases, technology improves the quality of teaching and learning, including basic skills such as digital literacy. However, it is important to recognize that it can also be exclusionary, irrelevant, costly, or harmful. UNESCO also stresses the importance of governments ensuring adequate conditions for equitable access to education, regulating the use of technology to protect students, and training teachers. The report recommends that technology should be introduced into education based on evidence of its suitability, equity, scalability, and sustainability. Implementation should priorities students' interests and complement education based on human interaction. (*Global Education Monitoring Report, 2023: Technology in Education: A Tool on Whose Terms?*, 2023).

One of the biggest barriers to digital transformation in higher education is resistance to change from teachers and administrators (Aditya et al., 2021, 2022; Galustov et al., 2021). It is vital that implementation takes place in close collaboration with these elements to avoid discomfort or lack of confidence in using new digital tools and methods (Bucăţa et al., 2022). Adequate preparation of teachers to effectively integrate technology into their teaching practice through digital training and capacity building is essential (Bucăţa et al., 2022; Gaebel & Morrisroe, 2023; Galustov et al., 2021; Quaicoe et al., 2023).

One of the main concerns is the excessive use of technology, especially by young people. In order to avoid overburdening students, it is essential to maintain a balance in learning by promoting pre-sessional interactions and non-technological activities. This not only contributes to students' mental health but also strengthens interpersonal relationships and social development (Ydyrysbayev et al., 2022).

To meet these challenges, it is essential that careful planning is undertaken to meet the needs of the students. The approach must be student-centered, prioritizing equal access and improving the quality of digital education (Bucăţa et al., 2022). This requires the active collaboration of all stakeholders in the educational process to ensure that the digital transformation of education is beneficial and equitable for all (Quaicoe et al., 2023).

Higher Education Institutions are strategically incorporating a variety of innovative technologies into their digital transformation initiatives. These technologies include Analytics, Cloud, Artificial Intelligence, MOOC, and Virtual and Augmented Reality (Alenezi, 2021; Fernández et al., 2023; Gaebel & Morrisroe, 2023; Holmström, 2022; INOUE et al., 2022; Mukul & Büyüközkan, 2023). Analytics play a key role in monitoring and tracking student progress, identifying different groups of students, and providing personalized teaching experiences and advice to students and staff. Cloud technologies are designed to ensure security and provide anytime, anywhere access. Artificial intelligence is improving student support by providing instant responses and streamlining administrative tasks in higher education (Barbosa et al., 2023; Oliveira & Pinto, 2023). In turn, MOOCs are being used to disrupt education by providing flexible and accessible learning opportunities (Alenezi et al., 2023). On the other hand, virtual and augmented reality are helping to create digital classrooms, enhancing understanding and engagement through immersive 3D learning environments. This holistic integration of cutting-edge technologies aims to enhance the educational experience, making it more personalized, efficient, and dynamic within the higher education landscape (Fernández et al., 2023; INOUE et al., 2022; Mukul & Büyüközkan, 2023). The Valencia Declaration 2023 affirms that "the university has a great capacity not only to contribute but also to lead sustainable development with rigor, critical thinking, and social commitment" (Declaración de Valencia: Universidad y Sociedad, 2023).

This paper presents the case of the Polytechnic Institute of Tomar as a higher education institution and how it has pursued digital transformation through its Laboratory of Pedagogical Innovation and Distance Education (LIED – *Laboratório de Inovação Pedagógica e Educação a Distância*).

LIED (http://lied.ipt.pt/) is an applied research laboratory that was created to support teachers in their digital transformation by promoting, strengthening, and diversifying teaching and learning in face-to-face and distance education through pedagogical innovation and information and communication technologies. The aims of LIED are to support the development of distance training in an innovative way by resorting to the use of information and communication technologies;

support and development of content for distance education and open online courses (MOOC – Massive Open Online Course); to promote research in distance education by collaborating with researchers from national and international institutions; to develop and take part in pedagogical innovation projects and news learning methodologies; to foster learning in a collaborative environment; to provide specialized solutions and services to the community in pedagogical innovation and distance education; to promote the development of teaching staff skills; and to promote research and training through scientific and technical events and publications.

LIED's activities are based on four strategic axes, namely: (1) Distance Learning; (2) Pedagogical Innovation; (3) Research and Cooperation; and (4) Services and Consultancy. Under the strategic axis "Research and Cooperation," LIED participates in various national and international projects and partnerships, including the STRONG project, which will serve as the basis for this article. The aim of the study is to present the process of creating digital educational content using an agile learning methodology, which is part of one of the activities that make up the STRONG project.

The structure of the chapter includes a section with a detailed description of the project under study, with its objectives and corresponding activities; a specific section on the creation of digital educational content, which presents the methodology adopted, the structuring process, the different phases of instructional design, and the identification of the methodology and tools used; a section with the results obtained in the process, according to the content produced, the teachers, the preparation, recording and editing; and finally a section with the final reflections on the research.

2 STRONG Project

The Polytechnic University of Tomar (IPT), as an institution of higher education with extensive experience, is deeply committed to promoting the development of information and communication technologies and the use of innovative digital tools.

The IPT recognizes the growing importance of the development of competences known as twenty-first-century skills, which play a crucial role in improving the quality of teaching provided by our teachers, according to DigCompEdu – a European framework for the digital competence of educators, a general reference framework to support the development of specific digital competences for educators in Europe (Redecker & Punie, 2017). The framework presents 22 competences, organized into 6 domains – such as Professional Engagement, Digital Resources, Teaching and Learning, Assessment, Empowering Learners, and Facilitating Learners' Digital Competence – and a progression model to help teachers assess and develop their digital competence (Redecker & Punie, 2017).

However, a major challenge to achieving the desired improvements is the lack of contextualized understanding of teaching practices and effective approaches to supporting the professional development of our teachers.

It is in this context that the STRONG – *Resilient Skills and Teachers Focused on the Next Generations* – project has been launched, with the main aim of empowering both teachers and students at this level of education, preparing them more comprehensively with resilience and digital skills essential for the modern labor market. This includes adapting to new ways of interacting and participating in an increasingly digital society to face the challenges of the COVID-19 pandemic and its consequences. This project is co-financed by POCH – Portugal 2020 Operational Program with the reference POCH-02-53I2-FSE-000010 and coordinated by LIED.IPT. The project was carried out over 18th months and strengthened the two initiatives under the PRR measures "Impulso Adultos" and "Impulso Jovens" STEAM, in which the IPT is a co-promoter, namely the A23 Polytechnic Network and the consortium "Entre o Tejo e o Mar" – CETM.

The STRONG project aims to achieve several key objectives, as defined by the IPT, namely:

- To promote innovation in the teaching and learning methods adopted by the IPT, in line with the practices of educational institutions internationally recognized for their innovative approaches.
- To facilitate teachers'access to tools and materials that provide students with practical skills relevant to the labor market, thus contributing to their professional readiness.
- To stimulate a culture of self-learning and collaboration between different stakeholders at different stages, including teachers, students, employers, and other educational institutions. This collaboration aims to enrich the learning environment and prepare students for the challenges of today.

The project consisted of 10 activities designed to achieve the objectives set for the IPT, each with different actions/tasks to be carried out:

- Activity 1 included the implementation of 2 diagnostic studies and the identification of new skills for future jobs.
- Activity 2 involved carrying out 1 diagnostic study on the impact of innovative teaching and learning practices for the future at the national and international levels.
- Activity 3 involved the organization of seminars on new skills for future jobs, emphasizing knowledge transfer, employability, and innovation.
- Activity 4 aimed to create a digital infrastructure for knowledge transfer that promotes employability through communication between higher education and business.
- Activity 5 pretended to organize a seminar to present the platform developed in Activity 4.
- Activity 6 included immersion and experimentation activities for employability and knowledge transfer, namely study visits and training activities.
- Activity 7 involved the production of digital content to promote knowledge transfer, employability, and lifelong learning. This activity is specifically addressed in this study.

- Activity 8 includes mentoring activities, in particular internships in enterprises.
- Activity 9 covered the dissemination of training provision at the national and international levels.
- Activity 10 concerned the follow-up, evaluation, and monitoring of the implementation of the project.

The STRONG project has adopted a comprehensive and integrated strategy aimed at empowering teachers and students to take an active role in society. This approach is designed to recognize and balance their social, family, and personal responsibilities and to recognize them as adults committed to the sustainable growth of their local communities.

3 Methodology

As part of the STRONG project (Activity 7), a group of IPT professors who teach the curricular units of the courses belonging to the A23 and Tejo&Mar consortium were invited to transform the content of their subjects/curricular units into digital educational content in MOOC format.

MOOCs, as the name suggests, are Massive Open Online Courses in a variety of fields that give students from all over the world access to high-quality courses, breaking down geographical and financial barriers (Carvalho et al., 2020). These courses are flexible in terms of scheduling and allow students to learn at their own pace, making them ideal for people who want to improve their knowledge and skills in their field of study or work while keeping up to date. One of the special features of MOOCs is their ability to incorporate innovative technologies and teaching methods to enhance the learning experience (Deng & Gao, 2023). For students, the benefits include access to certification from prestigious universities and the opportunity to take distance learning courses that they would not otherwise be able to attend. In addition, MOOCs can encourage collaboration between students from different parts of the world, fostering the creation of a network between students, sharing of knowledge, and exchange of experiences (Alturki & Aldraiweesh, 2023).

To do this, it was necessary to outline a development program based on the Agile methodology, with five phases (Fig. 1).

Traditionally associated with software development, the Agile methodology considers that the process should be "an incremental and iterative approach" (Salza et al., 2019). This means that throughout the process all elements are involved in finding the best solution for the objective they are working on. Its use in online education is becoming increasingly common because of the benefits it offers. This

Fig. 1 Online course development process

Table 1 Course development phases based on agile methodology

Phase	Duration	Description	Instruments
Analysis	1 month	Formation of the multidisciplinary team Creation of the templates	Templates for presentation
Design	1 month	Course design workshop Design template	
Development	2 months	Media production workshop Integration on the platform	
Implementation	1 month	Pilot phase Analysis Final launch	Satisfaction survey LMS users data
Assessment	2 to 4 months	Analyze performance data and student satisfaction Improvements	Final satisfaction survey LMS users data

approach emphasizes flexibility, collaboration, and incremental involvement, making it ideal for creating online courses that meet the changing needs of students.

Five phases were outlined for the creation and implementation of the courses, based on the agile methodology (Table 1). A team of instructional designers and content specialists was established to support the trainers throughout the process.

In the Analysis phase, a multidisciplinary team was formed including trainers, instructional designers, developers, and content specialists. During this phase, templates were created for the videos to be produced, with a graphic identity associated with the institution, the institutional logos, and a reference to the project. The templates were proposed by the team of instructional designers and validated by the coordinators of Activity 7, which is responsible for producing digital content.

In the Design phase, a course design workshop was held for the teachers of the curricular units identified and organized by the team of instructional designers together with the activity coordinators. The purpose of the session was to support the preparation of the content of the different courses, in MOOC format. The design model for all the courses was defined and validated by the multidisciplinary team.

In the Development phase, a media production workshop was organized for the teachers of the identified courses, led by a team of instructional designers. The aim of the session was to support the creation of multimedia and multi-format content. The session had the support and feedback of the team of instructional designers and content specialists.

The next phase is Implementation, where the courses are introduced to a pilot group of students to collect data on student performance and to assess the effectiveness of the initial modules. It is important to analyze the information and improve the content to respond to the difficulties identified. After that, the final courses will be launched for all students.

The final phase will be Assessment, where it will be possible to collect final feedback from students and analyze performance data and student satisfaction to measure the success of each MOOC. After this phase, it will be possible to make continuous improvements based on the feedback.

The NAU platform (National Platform for Hosting MOOCs) is developing a satisfaction questionnaire to be implemented in its courses. This questionnaire is based on previous experiences where satisfaction surveys have been developed (Marques et al., 2022). The aim is to create a questionnaire that standardizes the collection of data on the satisfaction of students participating in MOOCs, allowing for a better comparison of data. To this end, we will use the same questionnaire for the pilot test and for the final evaluation of the course.

The aim of the whole process was to train trainers to meet the demands of online teaching, with the emphasis on autonomous interaction between students and content.

4 Results

The role of the instructional designers was to support teachers design and develop online courses for each of their subjects.

The work carried out in Activity 7 involved a total of 16 teachers from different domains of the courses in Human Resources Management, Tourism Management, IT, Digital Marketing, Security and Civil Protection, Information Systems Technology and Programming, Maintenance, and Rehabilitation of Railway Systems and Pedagogical Innovation, and involved the transformation of 24 curricular units into online courses, such as:

- Twenty-First Century Skills for the Next Generations
- Administrative Human Resource Management
- Applied Mathematics
- Business Intelligence
- Computer Algorithms
- Digital Marketing
- E-commerce
- Excel in the Social Sciences and Humanities
- Gamification
- Gastronomic Heritage and Culture
- Information Systems and Support for Human Resources Management
- Internet Technologies
- Introduction to Management
- Labor Law
- Mobile Programming
- Multimedia Systems in Cultural Heritage
- Pedagogical Resilience in Higher Education
- Problem Solving, Algorithms and Programming
- Programming I
- Programming II
- Society and Economy

Table 2 Phases of the online course design and development process

Stages	Deliverables
Planning	Planning the stages of the process Pilot learning sequence
Course design	Workshop 1: Course design with learning sequences – definition of modules, learning objectives, activities, and assessment Teacher materials
Content production	Workshop 2: Media production – Slides, scripts, and videos Activities Other content
Course creation on the project platform	Workshop 3: Integration of content on the Moodle platform – Courses reviewed, tested, and built on the platform
Course launch	(Under development)

- Spatial planning, Management and Organization
- Web Programming
- Writing for the Digital Environment

The process of designing and developing the online course had five stages: (1) Planning, (2) course design, (3) content production, (4) course creation on the project platform, and (5) course launch (Table 2).

In the first planning stage, the instructional designers built the pilot learning sequence together with the team coordinating Activity 7.

In the second stage, the instructional designers dynamized a Workshop of Course Design, in the Onboarding Day event. This event is designed to gather teachers and support staff to explain the process, start designing courses, have a first contact with the studio recording, plan the work, and exchange ideas. After the onboarding day, the instructional designer worked with the teachers to finalize the course design, define the learning sequences, collect materials, and prepare for the production phase. The Course Design allowed teachers to plan each of their course units by defining learning objectives, modules and topics, activities and resources, assessment, and estimated workload. The first step was to structure the course into modules and topics, the second step was to define the learning objectives for each module/topic, and the third step was to define learning sequences for the modules/topics (Fig. 2).

During the course design workshop, an explanatory video was shown on how to define learning objectives using Bloom's Taxonomy (Fig. 3). Bloom's taxonomy is a structure for the hierarchical organization of educational objectives proposed by Benjamin Bloom and his associates in 1950. The taxonomy organizes the learning process into six levels: (1) remember (knowledge), (2) understand (comprehension), (3) apply (application), (4) analyze (analysis), (5) evaluate (evaluation), and (6) create (creation). The first level involves the ability to recall previously learned information, such as facts, terms, and concepts. The second level refers to the ability to understand the meaning of information, explaining ideas, or interpreting messages. The third level involves the ability to use knowledge in a practical way, applying it to new situations or contexts. The fourth level refers to the ability to break down

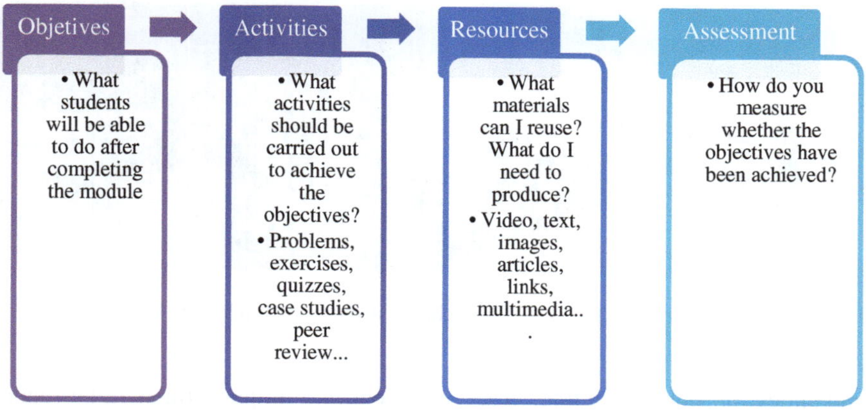

Fig. 2 Course design workshop

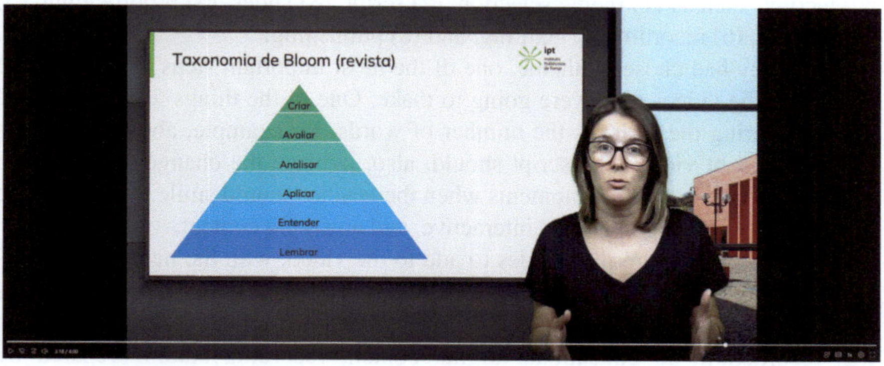

Fig. 3 Video on how to define learning objectives using Bloom's Taxonomy

information into parts and understand the relationship between these parts. The fifth level involves the ability to make judgments and evaluations, considering evidence and criteria. The sixth level, the highest level, refers to the ability to create something new by combining information in innovative ways, proposing solutions, or generating new ideas. Teachers can use Bloom's Taxonomy to plan learning activities, and assessments and to set expectations for student performance.

Teachers started the course design process for each of their subjects with the help and support of the instructional designers (Fig. 4).

In the third stage of Content production, the instructional designers promoted a Workshop of Media Production. The workshop aimed to support teachers in producing educational videos. The workshop started with some basic concepts about the differences between face-to-face and online teaching, and how the different processes take place. Based on the course design, which identifies all the activities and resources to be produced through learning sequences, they began to produce scripts

Fig. 4 Course design workshop in the Polytechnic University of Tomar

and slides for video recording, produce content in multiple formats (after recording), and create activities (exercises and tests).

The instructional designers then introduced the teachers to the 8 stages of video production, such as (1) theme selection, (2) script, (3) slides, (4) proofreading, (5) preparation, (6) recording, (7) editing, and (8) publishing.

Once they had chosen a theme, one of the most important steps was to prepare scripts for the videos they were going to make. One of the things to keep in mind when preparing the script is the number of words, for example, about 130 words equals 1 min of video. The script should, also, indicate the change of slides, the moments of pause, and the moments when the teacher should smile. All details are important for creating dynamic, interactive, and engaging content.

The teachers prepare their slides to add to the video, with the help of the team. Before the recording of the educational videos began, the instructional designers provided examples of good practice to help teachers record the content. The video was recorded in an educational digital content recording studio prepared by LIED.IPT with the necessary equipment (Fig. 5). Later, the team of instructional designers edited the videos, integrating the layout created previously.

In the fourth stage, the instructional designers organized a Workshop to integrate the digital content on the platform. The team created a Moodle platform with a customized template for the different curricular units (Figs. 6 and 7).

During the workshop, the teachers learned how to structure their courses (as we can see in Fig. 8), according to the modules and topics of the subjects, and how to integrate the different content (videos, documents, activities) to the platform according to the course design.

In this way, the teachers were able to build their courses around the content they produced (as we can see in Fig. 9, in the Internet Technologies course). The beta version of the courses has also been reviewed and tested.

The final stage of the work is the launch of online courses, which is currently under development.

The aim of this whole process is to equip teachers with the necessary tools they need to create engaging digital educational content. Following the work carried out, it was noted by the teachers involved in the whole process that the level of trust

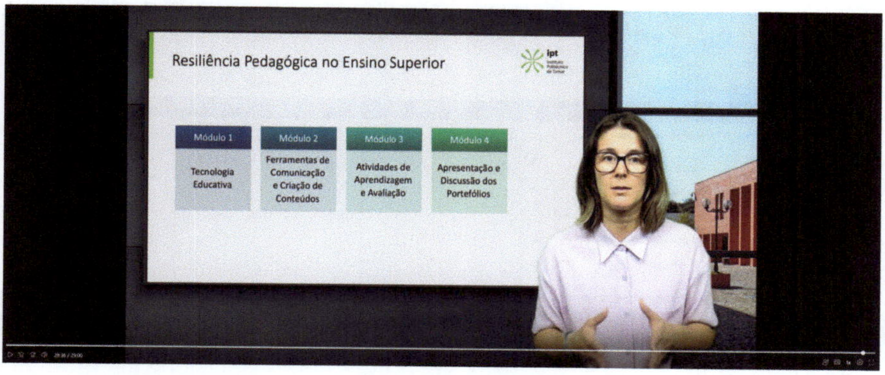

Fig. 5 Example of a video shot in the studio, with the slides and the institutional layout

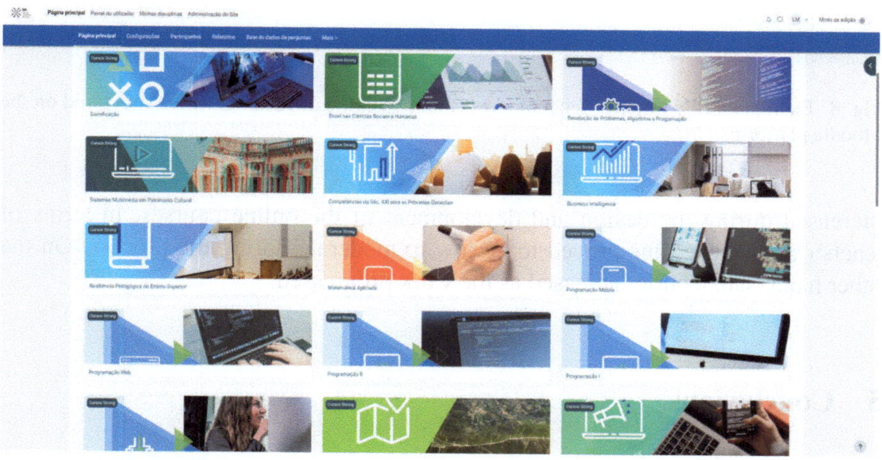

Fig. 6 Moodle platform with course integration using a personalized template

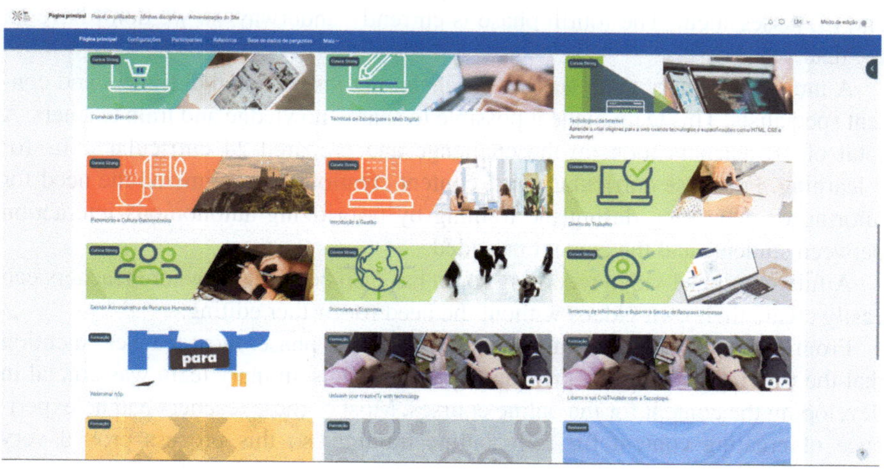

Fig. 7 Moodle platform

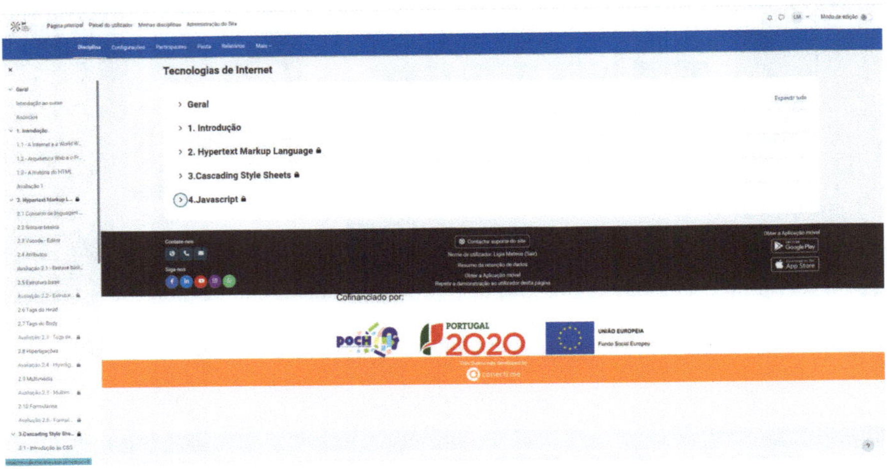

Fig. 8 Example of the Internet Technologies course with the different modules structured on the Moodle platform

increased during the design and development of the online courses. In terms of teacher attendance, this proved to be low to moderate during the process. On the other hand, satisfaction increased as the work progressed.

5 Conclusions

Through this project, it was possible to outline a process for training IPT teachers to develop educational digital content. The process took place in five phases, using the agile methodology: (1) analysis, (2) design, (3) development, (4) implementation, and (5) Assessment. The fourth phase is currently underway, so we don't have all the data yet.

A multidisciplinary team was set up, including instructional designers and content specialists. This team made it possible to share knowledge and train teachers. A total of 16 teachers took up the challenge and prepared 24 curricular units for e-learning. For these curricular units, content creation has minimized the need for tutoring or mentoring during the training by prioritizing autonomous interaction between students and the content provided.

A fully equipped video recording room has also been set up where teachers can easily create their own videos without the need for further editing.

From observations made during the development phase, most teachers mention that the training and support provided by the multidisciplinary team was crucial in developing the content for the online courses. Most of these teachers had no experience of creating content for fully online training, so this process proved very productive.

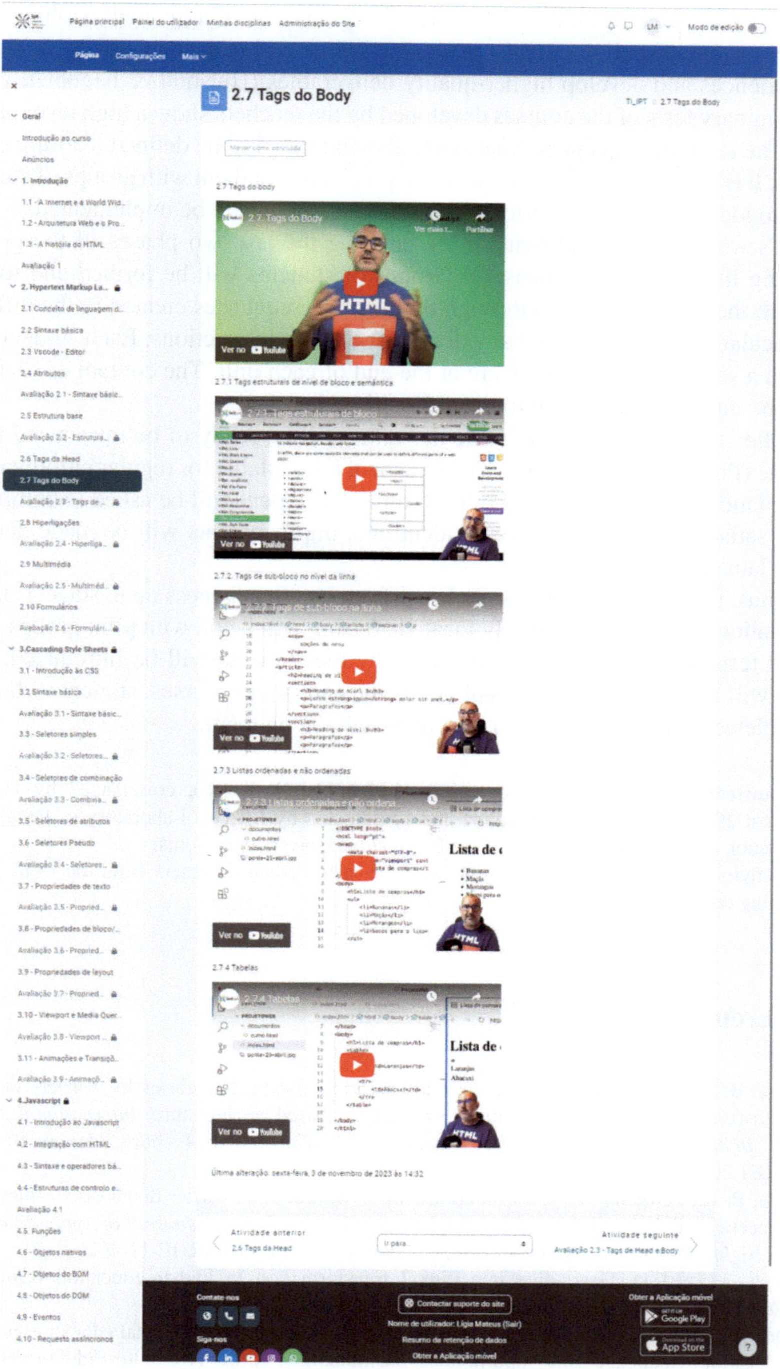

Fig. 9 Video content from Module 2, Topic 2.7, of the Internet Technologies course

According to Hulshult and Krehbiel (2019), the use of agile methodology in the development of online courses made it possible to build more effective learning experiences and develop higher-quality deliverables (Hulshult & Krehbiel, 2019). Preliminary tests of the courses developed by the teachers show a high level of clarity. The learning sequences created are also in line with the defined learning objectives. It is necessary to continue with the project to test them with groups of students and to identify possible shortcomings or improvements to be implemented.

It's with this in mind that we are entering the last two phases of the project. During Implementation (phase 4) Groups of students will be formed and for one month they will have to go through the learning sequences created in the different curricular units, pointing out any doubts or unclear instructions. Each student must fill in a satisfaction questionnaire at the end of each unit. The content of each unit will be improved according to the difficulties identified.

After the pilot study, the curricular units will be ready to be introduced to the public (Phase 5 – Assessment). The units will be available for regular enrolment and attendance periods. At the end of each unit, the student will be asked to complete a final satisfaction questionnaire. If identified, improvements will be made between each launch of the same unit.

Thus, the implemented process has proven its effectiveness up to stage 4 (implementation), where it is currently located. Testing will begin with pilot groups. After these tests, the improvements identified by the students will be introduced. Only then will the courses be made public. At the end of the courses, students who have completed them will be able to give their opinion on them.

Acknowledgments Project reference POCH-02-53I2-FSE-000010, co-financed by POCH - Portugal 2020 Operational Programme and coordinated by LIED – Laboratory of Pedagogical Innovation and Distance Learning, of the Polytechnic University of Tomar.

Activity carried out in collaboration with the instructional designers from the Delft Digital Learning company.

References

Aditya, B. R., Ferdiana, R., & Kusumawardani, S. S. (2021). Categories for barriers to digital transformation in higher education: An analysis based on literature. *International Journal of Information and Education Technology, 11*(12), 658–664. https://doi.org/10.18178/IJIET.2021.11.12.1578

Aditya, B. R., Ferdiana, R., & Kusumawardani, S. S. (2022). A barrier diagnostic framework in process of digital transformation in higher education institutions. *Journal of Applied Research in Higher Education, 14*(2), 749–761. https://doi.org/10.1108/JARHE-12-2020-0454

Alenezi, M. (2021). Deep dive into digital transformation in higher education institutions. *Education Sciences, 11*(12). https://doi.org/10.3390/educsci11120770

Alenezi, M., Wardat, S., & Akour, M. (2023). The need of integrating digital education in higher education: Challenges and opportunities. *Sustainability, 15*(6), 4782. https://doi.org/10.3390/su15064782

Alturki, U., & Aldraiweesh, A. (2023). An empirical investigation into students' actual use of MOOCs in Saudi Arabia higher education. *Sustainability, 15*(8), 6918. https://doi.org/10.3390/su15086918

Barbosa, L., Souza, P., Antonio, G., Joerke, O., Macedo, Y. M., Ferreira Vale, R., De Pádua, A., Oliveira, J., Suarez, M., Santo, D., Gomes, C. A., Cristina, S., Gomes, V., Alberti, R., Flávio, J., & Paz, D. (2023). Inteligência Artificial Na Educação: Rumo A Uma Aprendizagem Personalizada. *IOSR Journal Of Humanities And Social Science (IOSR-JHSS), 28*, 19–25. https://doi.org/10.9790/0837-2805031925

Bucăţa, G., Popescu, F., & Tileagă, C. (2022). Digital transformation of higher education system. *International Conference Knowledge-Based Organization, 28*(1), 158–168. https://doi.org/10.2478/kbo-2022-0025

de Carvalho, C. V., Butkiene, R., Gudoniene, D., Zubikova, O., & Bruegmann, T. (2020). Work in progress: A MOOC-based innovative instructional approach for curriculum design. *IEEE Global Engineering Education Conference (EDUCON), 2020*, 1844–1847. https://doi.org/10.1109/EDUCON45650.2020.9125394

Declaración de valencia: universidad y sociedad. (2023). https://www.Santander.Com/Content/Dam/Santander-Com/Es/Contenido-Paginas/Sala-De-Comunicaci%C3%B3n/Especiales/Encuentro-Internacional-De-Rectores-Universia/Valencia-2023/Do-Eiru-Valencia-Declaracion-2023-Es.Pdf

Deng, R., & Gao, Y. (2023). Using learner reviews to inform instructional video design in MOOCs. *Behavioral Sciences, 13*(4), 330. https://doi.org/10.3390/bs13040330

Ellström, D., Holtström, J., Berg, E., & Josefsson, C. (2022). Dynamic capabilities for digital transformation. *Journal of Strategy and Management, 15*(2), 272–286. https://doi.org/10.1108/JSMA-04-2021-0089

European Commission. Digital education action plan. (2018). https://Ec.Europa.Eu/Educa Tion/Education-in-the-Eu/Digital-Education-Action-Plan_en.

Feliciano-Cestero, M. M., Ameen, N., Kotabe, M., Paul, J., & Signoret, M. (2023). Is digital transformation threatened? A systematic literature review of the factors influencing firms' digital transformation and internationalization. *Journal of Business Research, 157*, 113546. https://doi.org/10.1016/J.JBUSRES.2022.113546

Fernández, A., Gómez, B., Binjaku, K., & Meçe, E. K. (2023). Digital transformation initiatives in higher education institutions: A multivocal literature review. *Education and Information Technologies., 28*, 12351. https://doi.org/10.1007/s10639-022-11544-0

Gaebel, M., & Morrisroe, A. (2023). The future of digitally enhanced learning and teaching in european higher education institutions. www.eua.eu

Galustov, R. A., Gerlach, I. V., Kopchenko, I. E., Spirina, O. N., Tvelova, I. A., & Khlopkova, V. M. (2021). Digital transformation of education and the analysis of possible risks: Results of a survey of teachers of general education organizations. *Perspektivy Nauki i Obrazovania, 53*(5), 446–462. https://doi.org/10.32744/PSE.2021.5.31

Global Education Monitoring Report. (2023). *2023: Technology in education: A tool on whose terms?* GEM Report UNESCO. https://doi.org/10.54676/UZQV8501

Hashim, M. A. M., Tlemsani, I., & Duncan Matthews, R. (2022). A sustainable university: Digital transformation and beyond. *Education and Information Technologies, 27*(7), 8961–8996. https://doi.org/10.1007/s10639-022-10968-y

Holmström, J. (2022). From AI to digital transformation: The AI readiness framework. *Business Horizons, 65*(3), 329–339. https://doi.org/10.1016/j.bushor.2021.03.006

Hulshult, A. R., & Krehbiel, T. C. (2019). Using eight agile practices in an online course to improve student learning and team project quality. *Journal of Higher Education Theory and Practice, 19*(3). https://doi.org/10.33423/jhetp.v19i3.2116

Inoue, M., Tsunoda, K., Nagahara, Y., Yaegashi, R., Ishizaki, H., & Maruyama, T. (2022). Digital transformation of higher education. *Journal of JSEE, 70*(3), 3_3–3_8. https://doi.org/10.4307/jsee.70.3_3

Konopik, J., Jahn, C., Schuster, T., Hoßbach, N., & Pflaum, A. (2022). Mastering the digital transformation through organizational capabilities: A conceptual framework. *Digital Business, 2*(2), 100019. https://doi.org/10.1016/j.digbus.2021.100019

Kraus, S., Jones, P., Kailer, N., Weinmann, A., Chaparro-Banegas, N., & Roig-Tierno, N. (2021). Digital transformation: An overview of the current state of the art of research. *SAGE Open, 11*(3). https://doi.org/10.1177/21582440211047576

Marks, A., AL-Ali, M., Atassi, R., Abualkishik, A. Z., & Rezgui, Y. (2020). Digital transformation in higher education: A framework for maturity assessment. *International Journal of Advanced Computer Science and Applications, 11*(12), 504–513. https://doi.org/10.14569/IJACSA.2020.0111261

Marques, C. G., Nata, A., Manso, A., & Araújo, I. (2022). Promoção do ensino aberto e flexível no Instituto Politécnico de Tomar: Projeto NAU. *Revista Da UI_IPSantarém, 10*(4), 5–17.

McCarthy, A. M., Maor, D., McConney, A., & Cavanaugh, C. (2023). Digital transformation in education: Critical components for leaders of system change. *Social Sciences and Humanities Open, 8*(1), 100479. https://doi.org/10.1016/j.ssaho.2023.100479

Mhlanga, D., & Moloi, T. (2020). COVID-19 and the digital transformation of education: What are we learning on 4ir in South Africa? *Education Sciences, 10*(7), 1–11. https://doi.org/10.3390/educsci10070180

Mukul, E., & Büyüközkan, G. (2023). Digital transformation in education: A systematic review of education 4.0. *Technological Forecasting and Social Change, 194*, 122664. https://doi.org/10.1016/j.techfore.2023.122664

Nadkarni, S., & Prügl, R. (2021). Digital transformation: A review, synthesis and opportunities for future research. *Management Review Quarterly, 71*(2), 233–341. https://doi.org/10.1007/s11301-020-00185-7

Nóvoa, A., Cirilo, P. R., Nascimento Silva, P., & Nonato, B. F. (2023). Desafios e perspectivas contemporâneas da docência universitária. *Revista Docência Do Ensino Superior, 13*, 1–20. https://doi.org/10.35699/2237-5864.2023.48009

OECD Skills Outlook 2023 (2023). OECD. https://doi.org/10.1787/27452f29-en

Oliveira, L., & Pinto, M. (2023). A Inteligência Artificial na Educação – Ameaças e oportunidades para o processo ensino-aprendizagem. https://www.esmad.ipp.pt

Quaicoe, J. S., Ogunyemi, A. A., & Bauters, M. L. (2023). School-based digital innovation challenges and way forward conversations about digital transformation in education. *Education Sciences, 13*(4). https://doi.org/10.3390/educsci13040344

Redecker, C., & Punie, Y. (2017). *Digital competence of educators DigCompEdu*. Publications Office of the European Union.

Safonov, Y., Usyk, V., & Bazhenkov, I. (2022). Digital transformations of education policy. *Baltic Journal of Economic Studies, 8*(2), 127–136. https://doi.org/10.30525/2256-0742/2022-8-2-127-136

Salza, P., Musmarra, P., & Ferrucci, F. (2019). Agile methodologies in education: A review. In *Agile and lean concepts for teaching and learning* (pp. 25–45). Springer Singapore. https://doi.org/10.1007/978-981-13-2751-3_2

Van Veldhoven, Z., & Vanthienen, J. (2022). Digital transformation as an interaction-driven perspective between business, society, and technology. *Electronic Markets, 32*(2), 629–644. https://doi.org/10.1007/s12525-021-00464-5

Ydyrysbayev, D., Kakimova, L. S., Sailaubaikyzy, B. G., Talgatbekovich, S. Y., Urmatova, A., & Orazbaev, E. (2022). Determining the digital transformation in education in the society 5.0 process. *International Journal of Emerging Technologies in Learning, 17*(18), 136–145. https://doi.org/10.3991/ijet.v17i18.32331

Public Policy and Digital Transformation in Higher Education

Welinton Baxto

1 Introduction

The Brazilian population is among the countries with the most access to the Internet,[1] although it is unequal in terms of the type of network and bandwidth (broad or narrow) for purposes of application in formal education environments. In the 2022 Census, applied by the Brazilian Institute of Geography and Statistics (IBGE, 2022), it was identified that the Brazilian population has 203,062,512 inhabitants, in a territorial area of 8,510,418 km^2 and an average demographic density of 23.86 inhabits./km^2. From these data, we can extract different clippings, whether the population density depends on the region, whether southeast or south, with higher density to the detriment of the north, northeast, or Midwest regions, with lower density in some states.

Comparatively, the 2022 Census obtained a growth rate of 0.52% higher than in previous years. According to IBGE (2022), the absolute variation and the geometric growth rate in the period 2010/2022 were obtained through the resident population in 2010 (reconstituted according to the territorial base of 2022) and the resident population in 2022. For this last year, the estimated population (approximately eight million inhabitants) was included for occupied permanent private households without an interview.

[1] Countries with the largest digital populations in the world as of January 2023. Available at (https://www.statista.com/statistics/262966/number-of-internet-users-in-selected-countries/)

W. Baxto (✉)
University of Brasilia, Brasília, Brazil

A. de Bem Machado et al. (eds.), *Digital Transformation in Higher Education Institutions*, EAI/Springer Innovations in Communication and Computing,
https://doi.org/10.1007/978-3-031-52296-3_4

Although data from *statista.com* point to a higher rate, 84.3% of Internet penetration in July 2023 for Brazil, studies presented by the Regional Center for Studies for the Development of the Information Society – Cetic.br,[2] Brazilian Internet Steering Committee (CGI.br, 2023), shows that "cultural equipment expands its online presence, but the Internet offer for the public is still reduced":

> […]
> The survey also shows that, in 2022, cultural facilities expanded their presence on the Internet through social networks or online platforms. The highest incidences were observed in movie theaters (from 85% in 2020 to 90% in 2022) and culture points (82% in 2020 to 89% in 2022). The presence was lower among libraries (37% had profiles on social networks or online platforms).
>
> The cultural institutions that most have their own websites are cinemas (75%), culture points (55%) and archives (60%). Movie theaters (29%) are the ones that provide the most applications for cell phones and *tablets,* followed by archives (13%).
>
> Among the activities most carried out by cultural facilities on websites and social networks, the following stand out: responding to comments and questions received – practiced by 83% of cinemas and 81% of culture points – and disclosing programming – an activity carried out by 85% of cinemas and 83% of the culture points. On the other hand, the least recurrent were: the dissemination of videos and audios or *podcasts* produced the sale of products and services and the realization of *lives* or online transmissions in real time. The sale of products and services through websites or social networks was the least performed activity by libraries (3%), museums (13%) and archives (15%).
>
> TIC Culture 2022 also investigated the offer of workshops or training activities for the public, either in person or online. Only heritage sites (3%), movie theaters (3%) and culture points (1%) reported offering exclusively remote access. Attendance only in person was reported in greater proportion by theaters (52%), culture points (45%) and museums (36%). Culture points (47%), archives (38%) and museums (30%) stood out for promoting proportionally more training activities in both face-to-face and distance learning activities. (Cetic.br, 2023)

The cultural events presented by Cetic.br are the most accessed, for example, online cinema. On the other hand, virtual libraries had less access in the period between 2020 and 2022. These references deserve close attention regarding the digital transformation of the population. Considering that in this period, the world was cornered by the pandemic caused by COVID-19,[3] putting pressure on several sensitive areas such as health, infrastructure, transport, and social area, which includes education. In that social confinement, the global economy was affected due to the sanitary measures enacted to contain the body-to-body contamination of the population.

In the educational area, the public authorities made use of emergency remote teaching in an attempt to provide educational assistance, in basic and higher education, by Ordinance No. 343, of March 17, 2020 (BRASIL, 2020a). Replacing face-to-face classes with classes in digital media, while the pandemic situation of the New Coronavirus – COVID-19 lasted. Authorizing, on an exceptional basis, the

---

[2] Cetic.br: Regional Center for Studies on the Development of the Information Society. Accessed at: https://www.cetic.br/pt/sobre/ and https://cetic.br/pt/pesquisa/cultura/indicadores/

[3] Source: Ministry of Health. Health Surveillance Secretariat (SVS): COVID19 Epidemiological Surveillance Guide. Panel developed to be the official communication vehicle on the epidemiological situation of COVID19 in Brazil. Available at: https://covid.saude.gov.br/

replacement of face-to-face subjects, in progress, by classes that use means and information and communication technologies (ICT), within the limits established by the legislation in force, by a higher education institution that is part of the federal education system, of dealing with art. 2 of Decree No. 9.235, of December 15, 2017 (BRAZIL, 2017c).

We have gone along decades with the application of ICT in basic and higher education, from undergraduate to postgraduate. However, we must not confuse remote teaching with the Distance Education modality (EaD).

Distance Education (EaD) was included in the law and basic guidelines by Law n.° 4,024, of December 20, 1961 (LDB 4,024/61), on an experimental basis, idealized by Darcy Ribeiro,

> Darcy Ribeiro, as he announced in a speech, compared himself to a snake with several skins Gomes (2010). A "naturalist", ethnologist, anthropologist, teacher, educator, politician, essayist and novelist, Darcy produced several works throughout his career and became a man of action. A man of many "skins", he was not an intellectual who was only thinking and writing. It was required to perform Gomes (2010). In this perspective, he idealized education as a path to social and national changes. (Amaro & Baxto, 2013)

Including, proposing at the time, the Open University of Brazil project, aiming at the democratization of education, launching efforts with other educators, the important aspects for EaD, mainly, the methodological style that approximated the practice of theory in this educational modality. As a modality, implemented by Law n° 9.394, of December 20, 1996, in its art. 80:

> The Government will encourage the development and dissemination of distance learning programs, at all levels and types of education, and continuing education. (Regulation)
>
> § 1 Distance education, organized with a special opening and regime, will be offered by institutions specifically accredited by the Union.
>
> § 2 The Union shall regulate the requirements for taking exams and registering diplomas related to distance education courses.
>
> § 3 The norms for the production, control and evaluation of distance education programs and the authorization for their implementation, will be up to the respective education systems, with the possibility of cooperation and integration between the different systems. (Regulation)
>
> § 4 Distance education will enjoy differentiated treatment, which will include:
>
> I – Reduced transmission costs in commercial sound and sound and image broadcasting channels and in other means of communication that are operated with authorization, concession or permission from the public authority; (Wording provided by Law No. 12,603 of 2012).
>
> II – Concession of channels with exclusively educational purposes;
>
> III – Reservation of a minimum time, free of charge to the Government, by concessionaires of commercial channels. (Law No. 9,394, 1996).

Based on these arguments, this chapter outlines higher education as a public policy and digital transformation, given the importance that this policy has developed. Since the 1990s, public authorities have been intensifying efforts in the distribution and training of teachers using ICT. Looking forward to the implementation of ICTs in teaching and learning processes. Actions aimed at improving educational quality, accessibility, inclusion, and the efficiency of educational processes.

Currently, these actions are related to raising the gross enrollment rate in higher education to 50% and the net rate to 33% of the population aged 18 to 24 years. In line with the National Education Plan (PNE, 2014–2024), approved by Law No. 13,005, of June 25, 2014.

Aiming to ensure the quality of the offer and expansion of at least 40% of new enrollments in the public segment. These guidelines aim, in addition to improving the quality of education, to promote humanistic, scientific, cultural, and technological development in the country. Reinforcing the need to achieve the twenty goals established in the PNE 2014–2024.

In this way, it is possible to observe some actions that were conceived by the federal government as a policy of digital transformation in national education, by the Ministry of Education (MEC) and its autarchies. For example, the Secretariat of Basic Education (SEB), Secretariat of Technological Education (SETEC), Secretariat of Higher Education (SESu), and Coordination for the Improvement of Higher Education Personnel (CAPES).

2 Digital Transformation Plan of the Ministry of Education

In this section, the general aspects of the Digital Transformation Plan[4]of the Ministry of Education are described. According to the federal government, the first step towards establishing the digital transformation strategy was identifying the universe of services offered by the federal executive branch. Subsequently, the Ministry of Education (BRASIL, 2023b) defined the unification of public services, covering government activities provided directly to users, through individualized transactions. The services identified in the initial survey were grouped on the single gov.br portal (https://www.gov.br/pt-br). The agency explained, at the time of the survey, which about a third of the services were already fully in digital format. Thus, migrating the existing structure did not cause major changes in programs and projects for the gov.br store. However, to enhance the digital transformation for citizens, companies, and government, as shown in Fig. 1:

The purpose of migrating to the single gov.br portal (https://www.gov.br/pt-br) was to improve service to citizens, reduce bureaucracy, simplify access, and reduce waiting times and costs. Costs for obtaining public services in the educational area are under its responsibility. For MEC (BRASIL, 2023b), this measure would translate into more efficiency, and speed, in addition to reducing infrastructure costs and optimizing the workforce.

In this regard, the aim was to simplify access to services, reducing costs and waiting times for citizens. It could also optimize workforce, infrastructure reduction, and delivery time.

[4]Digital Transformation, Ministry of Education. Accessed on 03 Jun. 2023. Available: https://www.gov.br/governodigital/pt-br/transformacao-digital/trajetoria-da-transformacao-digital

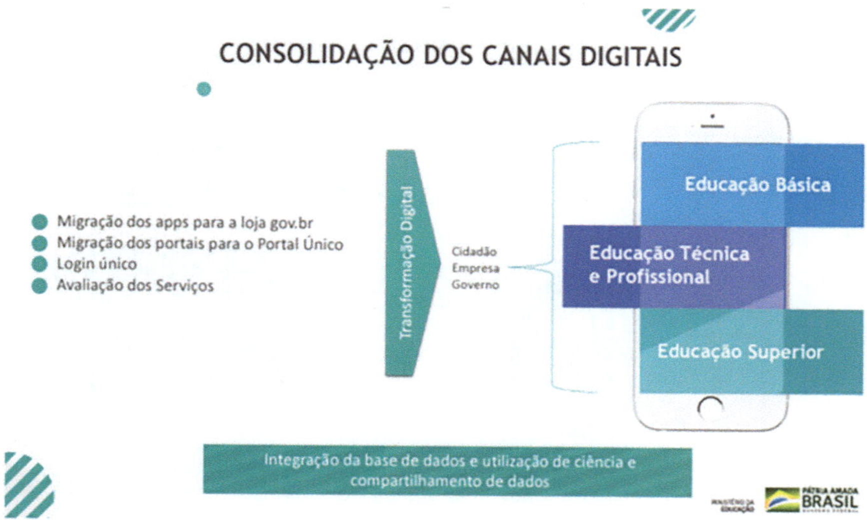

Fig. 1 Consolidation of digital channels. (*Source: Ministry of Education*)

These MEC movements seem to coincide with the low number of civil servants in direct administration in recent years. A fact that this reduction, in part, is due to the personnel policy in the different governments, especially, due to the absence of competitions to replace the workforce. Other factors that contributed to the drop in the workforce are due to pressure exerted by changes in retirement rules, or competition for other areas and powers (legislative and judicial) with more attractive plans and salaries.

It is not surprising that the agency sought the guidelines of the federal government to implement digital transformation in its structure. In other words, automation of educational services by direct and indirect administration. Not so much directed by the narratives to a possible improvement of the educational processes by the digital transformation, but in a clear urgency of the lack of effective public agents. A fact that there is an increase in contracts for consultancies and outsourcing in the body, justifying it due to the precariousness and lack of appreciation of the effective servants of the body (Fig. 2):

For the Ministry of Education (BRASIL, 2023b), some services would undergo digital transformation in higher education, basic education, and professional and technological education. This dimension would cover the services available at the Coordination for the Improvement of Higher Education Personnel (Capes), at the Brazilian Company of Hospital Services (Ebserh), at the National Education Development Fund (FNDE), and at the Anísio Teixeira National Institute of Educational Teaching and Research (Inep). These initiatives are ratified in five regulations. Decree 8936, of 2016 (2016b), which deals with the Digital Citizenship Platform. Decree 8638, of 2016 (BRAZIL, 2016a), on Digital Governance Strategy. Decree 9,723, of 2019, and Decree 9,094, of 2017 (BRAZIL, 2017b), which deal

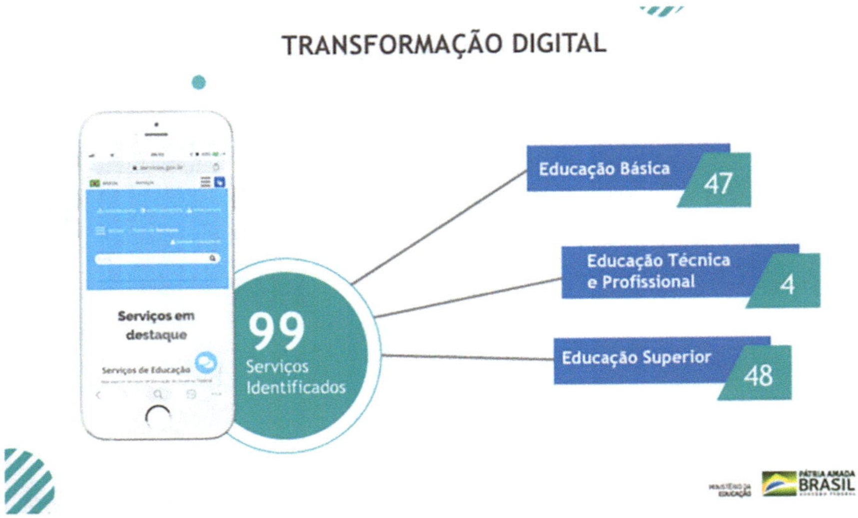

Fig. 2 Digital Transformation Ministry of Education. (*Source:* Ministry of Education)

with the simplification of services and the use of the CPF as the unique identification key of the citizen. Finally, Decree 9,756, of 2019 (BRAZIL, 2019b), which provides for the unification of government digital channels.

However, the digital transformation in higher education leads us to two programs, the Open University of Brazil (UaB) and Reuni Digital. Because they approach the characteristic of programs that favor the digital transformation in education. These programs are in line with Target 12 of the PNE 2014–2024, even initiated as government policies, pervade adverse ideologies, and remained as if they were, albeit precariously in some governments, as State policy.

3 Public Policy and Digital Transformation in Higher Education: Open University of Brazil (UaB) and Reuni Digital

From the perspective of public policy and digital transformation in higher education, two programs stood out, the Open University of Brazil (UaB) and Reuni Digital. Programs launched by the federal government, in distant decades, UaB in 2005 and Reuni Digital in 2022. However, the similarity is in the policy of expansion of higher education in the distance modality (EaD). The first is offered by public institutions (federal, and/or state, and/or municipal), the second by Federal Universities. The programs described, according to the public access source on the Ministry of Education portal (BRAZIL, 2023b).

3.1 Open University of Brazil (UaB)

We can cite as an example of public policy and digital transformation, due to the prerogative of expanding higher education, the UaB. Established by Decree No. 5,800, of June 8, 2006, (BRAZIL, 2006) as the Open University of Brazil (UaB) system, through the modality of Distance Education (EaD), although it is a ministerial program. It is considered that the UaB is used by the application of ICTs in the teaching and learning processes, in order to enable its wide use and modify the ways of teaching and learning, when compared to the last century. In other words, it may occur differently in environments immersed in digital technologies, depending on the infrastructure, initial and continuing training of education professionals, and the regional location of the higher education institution (HEI).

The EaD was included in Brazilian education by Law No. 4024, of December 20, 1961 (BRAZIL, 1961) on an experimental basis, idealized by Darcy Ribeiro, who envisioned the project of creating an Open University of Brazil (UaB). With a view to the democratization of education, conceiving efforts with other educators, for the important aspects of EaD. Emphasis on the methodological style that approximated the practice of theory in this educational modality. It is considered that it was Darcy Ribeiro who, with a broad vision of the future and always concerned with the democratization of access to education, even more than 30 years later, removed distance education from clandestinely.

Contrary to what is propagated in some media, DE is not so recent here and in the world. The first experiences date from 1833 in Sweden, 1840 in England, 1856 in Germany, 1874 in the United States, and, from 1974 onwards, in Pakistan, Thailand, Sri Lanka, Indonesia, and India. These countries use EaD to meet different objectives, for example, the consulting system, teacher training, and higher education, among others, and the most used mode of distribution for these courses was correspondence. In Brazil, since 1904, international schools have offered correspondence courses, according to Silva (2007, p. 24).

According to Mundim (2006, p. 119), "historically, distance education has evolved through different generations, following the technological development of telecommunications, information technology and the Internet." For the author, this development occurred in four generations, and in the first generation (1840–1970), courses were essentially carried out by correspondence with study guides and printed materials. In the second generation (1970–1980) with the framework of the Open Universities, the main means of distributing content began to carry out by radio and television, with print media used as complementary material. In the third generation (1980–1990), the use of video cassettes and satellite communication revealed a new way of presenting content through television, offering quality materials and allowing the study to be out at any time. In the fourth generation, computers and multimedia resources provide students with a higher level of interactivity and allow teachers to carry out pedagogical mediation through a virtual learning environment with synchronous (real-time) and asynchronous (independent time) communication modes. The fifth generation is on the intelligent flexible learning model; it is similar to the fourth generation since it is based on the use of the computer mediating communication. However, it differs from the fourth generation by showing greater economy of scale in teaching administration, academic support associated with an automatic response system. It also has a rich source of direct interaction, organized in databases and subsequently made available in databases in automated response systems.

In the face-to-face model, interaction occurs in real time, usually with synchronous communication. While, in the distance model, it can occur by synchronous and asynchronous communication. It is through this interaction that professor and student can dialogue, debate, share experience, receive guidance, overcome challenges, and encourage dialogue between theory and practice. It is in this context that the organization of pedagogical work becomes a significant element in educational practice in the virtual classroom.

In Brazil, the integration of ICTs in educational environments has, over the course of a few decades, materialized as digital technologies evolve. These growing integrations reveal some practices in initial and continuing education courses in basic and higher education institutions. Notably, after the regulation of art. 80 of Law No. 9,394, of December 20, 1996 (BRAZIL, 1996), by the then Decree No. 5,622, of December 19, 2005 (BRAZIL, 2005), revoked by Decree No. 9,057, of 25 of May 2017 (BRAZIL, 2017a), which regulated distance education (EaD), as an educational modality.

The importance of EaD in higher education in the country, as a possibility for digital transformation in higher education, among other aspects, is reflected in the growing evolution of the number of vacancies in the last two decades, suggesting that the modality has won its place in national education. That is, favored by the induction of public power, by the distance modality, via the UaB "System" and the expansion of private higher education institutions (HEIs).

In mid-2005, within the framework of the forum for state-owned companies for education, discussions took place towards the creation of the Open University of Brazil (UaB) in the distance education format. This would be under the responsibility of public institutions of higher education (IPES). Aiming to bring quality public higher education to those municipalities that did not have the offer of higher education or insufficiently offered to citizens. The political situation at that time made it possible to approve the UaB, with the benefit of offering a higher education course in the distance modality, with assistance to students at federal, state, or municipal on-site support centers.

According to the Higher Education Census 2021 (BRAZIL, 2021a, b), published with a time lapse of two years, by the National Institute of Educational Studies and Research Anísio Teixeira (Inep), the distance learning modality has grown. Currently, 62.8% of entrants in undergraduate courses, when compared to the face-to-face modality, expressed in graph 6 of INEP (Fig. 3):

According to INEP, the increase in the number of entrants between 2020 and 2021 is exclusively due to the distance mode, which had a positive variation of 23.3% between these years. In face-to-face courses, there was a decrease of −16.5%. Between 2011 and 2021, the number of admissions varied negatively (−23.4%) in

Fig. 3 Number of enrollments in undergraduate courses, by type of teaching – 2011–2021. (*Source:* Prepared by Deed/Inep based on data from the Higher Education Census)

face-to-face undergraduate courses. On the other hand, distance courses increased by 474.0%, the percentage participation of students entering undergraduate distance courses in 2011 was 18.4%. This participation, in 2021, was 62.8%. (BRAZIL 2021a, b).

In the same 2021 Census, he pointed out some challenges to accelerate the pace and direction of the expansion of higher education in line with the PNE 2014–2024 (BRAZIL, 2014). Among them, the increase in the gross enrollment rate in higher education to 50% and the net rate to 33% of the population aged 18–24 years. Thus, ensuring the quality of the offer and expansion to at least 40% of new enrollments in the public segment. It is possible to extract from the 2021 Census that the growth in the gross enrollment rate in education, in particular, is due to the distance modality (Fig. 4):

Note that the participation of public institutions is modest compared to the participation of private institutions. Even though it was referenced in the PNE 2014–2024, on the indication of strategies aimed at the use of means and ICTs in the teaching and learning process, with differentiated methodologies, resources, and distance education technologies that are supported by the Law of Guidelines and Bases n° 9394 of 1996.

The UNESCO (2015) already pointed out at the time that Brazil should improve the skills of teachers regarding the use of ICT in education. Result of research that showed large numbers of students, from different educational levels, had learning disabilities in different disciplines. The low absorption of scientific concepts could harm the inclusion of these individuals in modern society.

Thus, the provision of higher education in the distance mode would necessarily imply the guarantee of rights set out in the Federal Constitution of Brazil of 1988 (BRAZIL, 1988), set out in art. 206, VII, provision of education with the guarantee of quality standards. In this trajectory, it would be necessary for the HEI maintainers

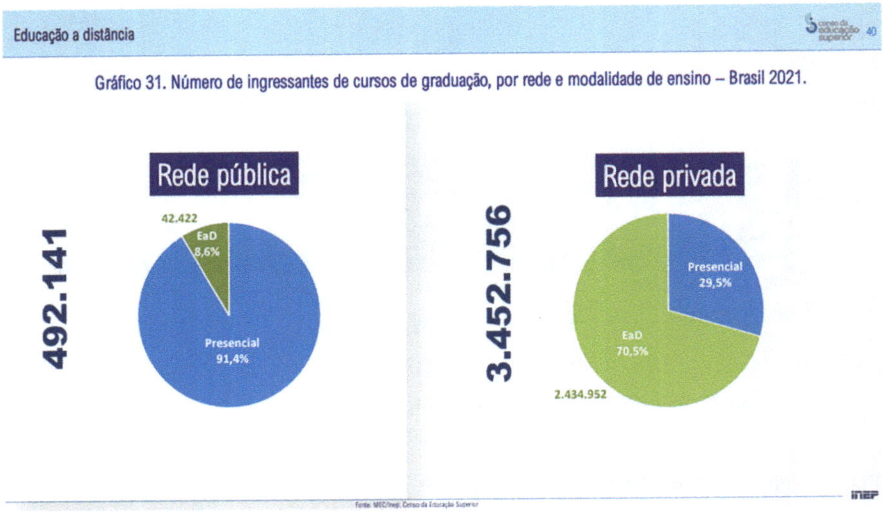

Fig. 4 Number of new entrants to undergraduate courses, by network and teaching modality – Brazil 2021. (*Source:* MEC/Inep; Higher Education Census 2021)

(public power and private initiative) to commit financial resources, and infrastructure, suited to the organizational model, for the application of digital technologies. Crucial points to raise the quality of the teaching and learning process, also in the courses offered in the distance modality.

That said, despite the higher education expansion policy in the country, described in the PNE 2014–2024, it is observed that public institutions, although considered a small network when compared to the private network, collaborated with 8.6% of the total of 492,141 enrollments. Reinforcing that, 35.6% of enrollments in undergraduate courses registered in the 2021 Census were offered by public institutions and 64.4% by private HEIs. Regarding the teaching modality, enrollments in face-to-face teaching courses represented 39.0%, while the distance modality accounted for 61.0% of total enrollments (BRAZIL, 2022).

Considering the purpose of the policy to expand higher education in the country, the 2021 Census distinguishes the distribution of students enrolled in undergraduate courses, see Fig. 5:

It is notable that 80% of students at public HEIs attend face-to-face courses. Since, in private HEIs, distance courses represent 84% of students. However, it is questioned whether these data are linked to the interests of students or is it a management policy of public HEIs in maintaining on-site courses to the detriment of distance courses.

The public HEIs, regardless of the expansion policy designed by the government in the distance modality, have low implementation. This leads us to low expansion, directly affecting the digital transformation of higher education, reverberating in the achievement of goal 12 (twelve) of the PNE 2014–2024.

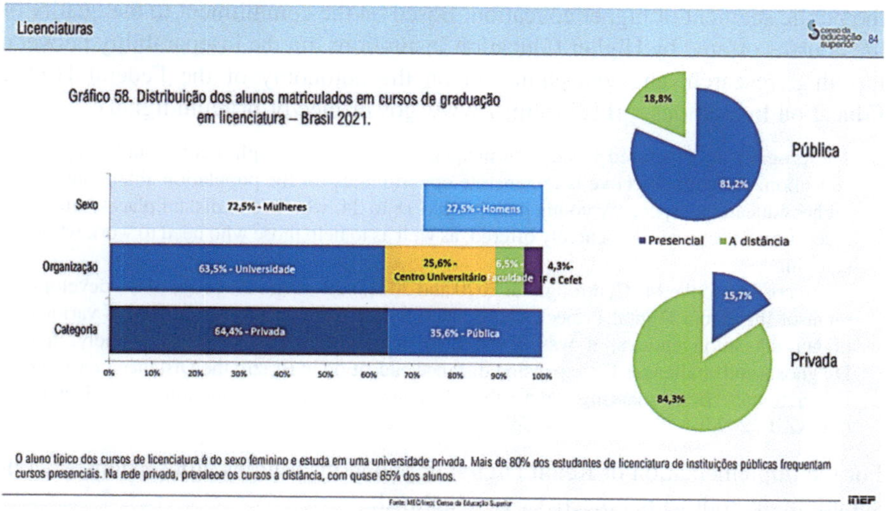

Fig. 5 Students enrolled in undergraduate degree courses – Brazil 2021. (*Source:* MEC/Inep; Higher Education Census)

3.2 Digital Reunion

As for the Reuni Digital project, according to MEC (BRAZIL, 2023b), it intends to expand, restructure, and optimize the offer of distance education through the network of Federal Institutions of Higher Education (IFES). In this context, the Strategic Guidance Committee (COE) established, established by Ordinance No. 433, of October 22, 2020 (BRASIL, 2020b), and the Working Group (GT), established by Ordinance No. 434, of October 22, 2020 (BRAZIL, 2020c). Committee and working group with the aim of subsidizing the discussion, elaboration, and presentation of strategies for expanding the offer of higher education courses, in the EaD modality, in federal universities. The Second COE (BRAZIL, 2020b) and the WG (BRAZIL, 2020c) are responsible for preparing a Plan for the Expansion of Higher Education through digital means. Among other expectations, the project should be delivered with an overview of EaD in the country, through a strategic analysis of international best practices (benchmarking*)* and a survey of the challenges of Higher Education in Brazil, in the distance modality.

According to the Ministry of Education (BRAZIL, 2023b), as a result, a scope was prepared to meet the Digital Transformation Plan, which provides for the migration of MEC applications to the Gov.br store and the portals to the single portal. Still, with the integration of the database, the use of science and data sharing will increase the efficiency of public policies that are based on evidence and, thus, will make decisions more objective and transparent.

Reuni Digital is structured in a program aimed at the expansion of Distance Education (EaD) in Federal Universities, promoting an increase in enrollments in the public segment of higher education. Based on the commitment to the quality of the courses offered by Higher Education Institutions, on the inseparability between teaching, research, and extension and on the autonomy of the Federal Higher Education Institutions – IFES (https://www.gov.br/mec/pt-br/reunidigital):

> The program is structured to address inequality in access to high-quality public higher education. Its main objective is to generate opportunities for the population able to attend higher education, especially young people aged 18 to 24, who live in distant places, where face-to-face vacancies are scarcely offered, as well as to help those who need to work while studying.
>
> According to the MEC, throughout 2020 and 2021, still within the scope of the development of the Reuni Digital Project, coordinated by the MEC, with the support of various higher education entities, it was possible to achieve relevant deliveries, namely: the Diagnosis and challenges for expansion distance education in Brazil; the Distance Learning International Benchmarking and the Distance Learning Panorama in Brazil. (BRAZIL, 2023b)

For the implementation of Reuni Digital in public institutions, after extensive consultation, the following objectives were defined:

> expand access and quality training to higher education courses for the population; contribute to the inclusion policy by helping students with socioeconomic vulnerability, physical disabilities and minority segments of the population to enter and remain; make possible the flexibility of schedules, potentializing to reduce the evasion of higher education; enhance the use and adequacy of the existing infrastructure of HEIs, to offer new vacancies in higher education courses; provide students of face-to-face courses with distance learning courses, contributing to their training during the length of stay established in the pedagogical project of their course; to contribute, through Distance Learning (EaD), to the dissemination, improvement and creation of Information and Communication Technologies (ICT), especially in federal universities. (BRAZIL, 2023b)

Furthermore, these actions resulted in the development of an EaD Expansion Plan in Federal Universities and another Monitoring Plan, the latter still in the process of elaboration. As for the implementation and monitoring by the MEC, there are many challenges for the digital transformation of the Brazilian educational system into a digital government that deserves special attention from its managers, supervisory bodies, and organized civil society.

4 Challenges for the Implementation of Digital Transformation in the Brazilian Education System in Digital Government

This approach is based on the description of the technical report of the Federal Court of Auditors (TCU), JUDGMENT 2487/2022 – PLENARY, TC 043.945/2021-0 (BRAZIL, 2022), on the maturity of the Brazilian Educational System in Digital Government, affection for MEC, universities, federal institutes of education, state and municipal education networks.

The TCU is the body that analyzes and judges the accounts of federal public resource administrators. On the other hand, the MEC is the highest body with powers to plan, guide, and coordinate, at the national level, the process of formulating policies for basic and higher education, instituted by Art. 11, II, a), combined with the Art. 20, I, of the Decree on Regimental Structure, Commission Positions and Trust Functions of the MEC (Decree No. 10.195, of December 30, 2019).

This technical document states that the MEC evaluated the maturity of its institution in Digital Government, which considered a positive initiative to understand the current stage of maturity of the body and to plan, guide, and coordinate the elevation of its level together with the level of maturity of the educational ecosystem (Brazilian educational system).

In this context, the Digital Transformation Plan was prepared with initiatives in the transformation of services, unification of digital channels, interoperability, security, and privacy. Depending on the plan, the special project "Learning Network + Student Journey" stands out. The construction of this platform was aligned with the National Education Plan of Law n.° 13.005, of June 25, 2014 (PNE 2014–2024), and the Guidelines and Bases of National Education of Law n.° 9.394, of 20 of December 1996 (LDB 9394/96). Platform designed to meet the Education Strategy and the National Policy for the Modernization of the State of Decree No. 10,609, of January 26, 2021. Regulates the Law on Access to Information (LAI) and Law No. 13,709, of August 14, 2018, General Law for the Protection of Personal Data (LGPD).

According to the MEC Digital Transformation Plan, Rede Aprendiz is an Integrated Information Platform on Brazilian Education, aiming to contribute to the evaluation and monitoring of educational public policies in Brazil.

In addition, the idea is to promote digital transformation in education, maintaining the priority given to student monitoring by educational institutions that play a decisive role in student life and should have their work ennobled with this system. It should be noted that the elimination of repetitive work with the automation and release of a new channel allows education professionals to act more effectively in direct support to the student, so that he has access to his information in a transparent way in a single channel, as shown in Fig. 6:

At this point, in JUDGMENT 2487/2022 – PLENARY, TC 043.945/2021-0 causes indicated about probable incipient maturity in Digital Government of Education (BRAZIL, 2022, p. 21) (Fig. 7):

Convergent with the Federal Court of Auditors (TCU), the digital education strategy for Brazil, the National Policy on Information and Informatics in Education, the Plan of Action, Monitoring and Evaluation of the Digital Education Strategy for Brazil, and the Management Committee of the Digital Education Strategy.

Along these lines, in Judgment 2487/2022 – plenary, TC 043.945/2021 – 0 (BRAZIL, 2022), it was observed that there is an indication in that inspection that the MEC had initiated actions associated with the Digital Government. However, he warned that success in this type of initiative is complex, as it depends on several other actors, including states and municipalities, with the autonomy of the Brazilian federative pact. In clear evidence that without articulation and engagement of the federal entities, the expected results not be achieved.

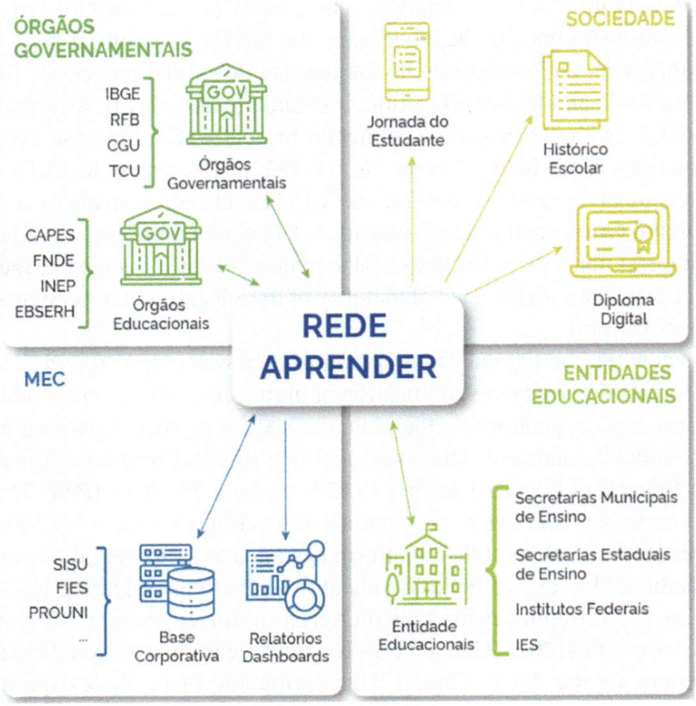

Fig. 6 Learning network. (*Source:* Ministry of Education)

Fig. 7 Digital government maturity levels in education. (*Source:* Elaborated by the Data Center in Education of the TCU)

In the same document, the TCU recommended efforts to involve and orchestrate the engagement of MEC public educational policy actors, such as universities, federal education institutes, municipal, state, and district education networks, other federal agencies, and private entities. In order to reach this conclusion, the TCU inspection team, TC 043.945/2021 – 0, took the following actions:

[…]

(a) Integrated analysis of the Basic Education Census and Higher Education Census databases;
(b) Analysis of the Managerial Treasury Basis. The purpose was to identify budget execution values for Digital Government actions in the Ministry of Health and Ministry of Economy to compare them with MEC values;
(c) Analysis of the Digital Transformation Data Analysis Panel of the Ministry of Education. The purpose was to identify the status of MEC digital government actions;
(d) Analysis of the Portal http://gov.br. The purpose was to identify the status of services and data for basic and higher education that the MEC has already unified on the portal;
(e) Comparative analysis (benchmarking) of Digital Health and the current situation of Digital Government in Education; It is.
(f) Documentary analysis of the MEC Information Technology Master Plan.
Results. (BRAZIL, 2022, p. 22).

In this effort, the TCU inspection team (BRAZIL, 2022) pointed out that the main finding refers to the incipient level of maturity in Digital Government of Education, and the causes are in the evidence of the situation found during the inspection:

- Difficulties of municipal and state networks to hire educational solutions or ICT solutions for the Digital Transformation (TD) of education;
- Lack of specialized workforce for TD projects;
- Lack of diagnoses of the federal, state and municipal education networks in DT and Digital Government of Education, as well as work processes for forwarding how this maturity can be increased, such as the induction, for educational managers, of the "10 steps for the digital transformation in states and municipalities," a practical guide to implement a digital strategy for the transformation of public services at the local level, or another selected by the MEC;
- Incipience of Digital Government mechanisms, such as national policy and updated and specific plans on the subject of Digital Government in Education;
- Inexistence of a governance instance with the participation of municipal, state and federal education networks formally constituted to deal with digital government and education TD;
- Need for improvement in digital skills of students, teachers and educational managers of municipal, state and federal networks. (BRAZIL, 2022, p. 22).

Law 14.129/2021 (BRASIL, 2021b), referring to the framework of the digital government of education, involves compliance with some points. Among them, the provision of digital services in public education such as enrollment, assessments, consultation of grades, when appropriate: digital books, online classes, face-to-face classes with digital technologies. Digitization of the MEC, the National Education Development Fund (FNDE), Inep, higher education institutions, other institutions within the scope of the MEC, and schools, municipal education departments, and state education departments of municipalities and states who adopt this law by their own normative act.

However, the Information Technology Inspection Secretariat (Sefti), of the TCU, in JUDGMENT 2487/2022 – PLENARY, presented recommendations to the Ministry of Education, as follows:

> […]
>
> II. Recommend to the Ministry of Education, based on art. 43, item I, of Law 8443/1992, combined with art. 250, item III, of the Internal Regulations of the Federal Court of Auditors, which:
>
> The. promote training and awareness actions on the Digital Government of education, digital transformation of education and digital skills aimed at managers of municipal, state and federal education networks, teachers, civil servants and service providers of schools and educational institutions, and students, with the aim of establishing a culture of Digital Government of Education, digitally transforming Brazilian education and developing digital skills for the 21st Century;
>
> B. carry out a diagnosis in Digital Government and Digital Transformation of Education with municipal, state and federal education networks, based on a maturity model in Digital Government, such as the model indicated in the PDTI 2021–2023 of the MEC or another chosen by the MEC, of in order to identify the maturity level of each surveyed network, contemplating the information exemplified in Annex 1 of Appendix G;
>
> w. based on the diagnosis carried out, induce actions to increase the level of maturity in Digital Government of the federal, state and municipal education networks, including, for example, forwarding guidelines on how the education network can reach higher levels of maturity in Digital Government ;
>
> d. prepare a single, consolidated document that integrates the Digital Education Strategy for Brazil; Plan of Action, Monitoring and Evaluation of the Digital Education Strategy for Brazil; and National Policy on Information and Informatics in Education; and that this single document defines, among others, the items exemplified in Annex 2 of Appendix G;
>
> It is. in conjunction with the Ministry of Economy, seek solutions to supply a workforce specialized in Digital Transformation, such as allocating Information Technology Analyst servers or using the Startup Gov.br Program, instituted by Ordinance SGD/ME 2,496, of 2 March 2021, as well as seek to strengthen the competence of MEC units in Digital Government and Digital Transformation, as support for the development of work processes and projects related to Digital Government and Digital Transformation of educational services and policies; and f. based on the constitutional principle of efficiency, assess the convenience and opportunity of adopting measures to attribute competence to an administrative body, within the scope of the Ministry, which is able, with the support of the Ministry of Economy, where appropriate, to influence priorities and projects of the business units regarding digital government and to manage Brazil's Digital Education Strategy. (BRAZIL, 2022, p. 28)

There is an effort by the public authorities to comply with TCU recommendations, JUDGMENT 2487/2022 – PLENARY, related to strengthening the competences of the MEC units, in Digital Government and in Digital Transformation. MEC has been engendering efforts with some strategies distributed among its Secretariats. The Secretary of Higher Education (SESu) was responsible for higher education programs and projects, some in partnerships. For example, Connected Students Program; IFES Monitoring Portal; Eduplay; Digital Diploma; Icpedu – National Digital Certificate Program; 360° University Platform; Platform For – Strategic Planning and Risk Management; Combating Evasion – Sissa Platform; Single

Access; Sisfundapoio. These programs and projects are described according to an excerpt extracted from the MEC website (https://www.gov.br/mec/pt-br) (BRAZIL, 2023a, b. WEB).

4.1 Connected Students Program

The Connected Students Program with actions by the Higher Education Secretariat (SESu) and the Professional and Technological Education Secretariat (Setec) of the Ministry of Education (MEC) allows the offer of public Internet access policies aimed at Federal Education Institutions and, above all, to students in conditions of socioeconomic vulnerability. It aims to guarantee the conditions for offering remote classes, through information and communication technologies (ICT) and the maintenance of academic activities during the duration of the social isolation measures affected by COVID-19. The Program enables connectivity for students and allows access to the content of classes remotely, thus mitigating the impacts during the period of social distancing caused by the COVID-19 pandemic, minimizing evasion, reducing the quality of teaching, and allowing digital inclusion.

4.2 Monitoring Portal of Federal Institutions of Higher Education (IFES)

The monitoring portal of Federal Institutions of Higher Education (IFES), under the responsibility of SESu/MEC, established the development of the Monitoring Portal of Teaching Institutions of the Federal Education Network. The action started during the pandemic, in partnership with the Federal University of West Bahia (UFOB), the Federal University of Cariri (UFCA), the Federal University of West Pará (UFOPA), and the Federal University of Viçosa (UFV). Its purpose is to monitor the operating status of Federal Universities and Federal Institutes and monitor the resumption of face-to-face academic activities in all educational institutions of the Federal Education Network. Also, publicize the main coping actions in the Federal Education Network and monitor the number of doctors, nurses, pharmacists, and physiotherapists who had their graduation early.

4.3 Eduplay

The Eduplay service was developed by the research laboratories of the Federal University of Paraíba (UFPB) and the University of São Paulo (USP). Partnership between MEC and the National Education and Research Network (RNP). The

Eduplay service allows access to live broadcasts, TV programs, exclusive channels, video lessons, access to radio stations, and *podcasts*. In addition, the Eduplay service has a network of channels that provide content in Brazilian Sign Language (Libra), as well as providing a network of varied programming, both with a focus on content of a university nature and educational content.

4.4 Digital Diploma

The Digital Diploma brings innovation, in the academic environment, which enables the modernization of the procedural flow for issuing and registering a diploma in Higher Education Institutions (HEIs) of the Federal Education System. It aims to be a document, with its existence, issue, and storage entirely in the digital environment. The adoption of the digital means for issuing diplomas must comply with the digital certification guidelines of the Brazilian Public Key Infrastructure standard – ICP-Brazil. The same for the National Validator refers to the verification and authenticity portal of the digital diploma issued by educational institutions.

4.5 National Digital Certificate Program (ICPEDU)

The National Digital Certificate Program with ICPEdu (Public Key Infrastructure for Teaching and Research) of MEC is a service offered in partnership with the National Teaching and Research Network (RNP). Enables the unlimited issuance of personal digital certificates for students, professors, and employees of teaching and research institutions, in a simple and freeway, since its main objective is to offer a virtual identity in a safe and agile way in academic processes and systems. The system is integrated with the Academic Identity Federation, the CAFe network, which already includes all Federal Universities and Federal Institutes of MEC and Research Institutes linked to the Ministry of Science and Technology.

4.6 360° University Platform

The University 360° platform, considered an observatory of Higher Education, was developed in partnership between the Secretariat of Higher Education (Sesu) and the Data Science BR Network – DSBR (research network in data science discovered by the Federal University of Western Bahia (UFOB). In cooperation with researchers from UFOPA, Federal University of Viçosa (UFV), University of Pernambuco (UFPE), and Federal University of Espírito Santo (UFES). The tool provides data transparency through integrated information from Federal Institutions of Higher Education (IFES), availability of academic and budgetary indicators, people management at Federal Universities, and TCU performance indicators.

4.7 *For Platform – Strategic Planning and Risk Management*

The For Platform (integration between ForPDI[5]and ForRisco[6]) is a technological application developed through researchers, specialists, and academics in the Information Technology (IT) area, which aims to meet the particularities of Higher Education Institutions, to encourage good strategic planning practices and institutional risk management. The system is designed to provide greater transparency and efficiency in the follow-up process of institutional planning, as well as in risk management at Universities and Federal Institutes.

4.8 *Fighting Evasion – SISSA Platform*

The SISSA Platform, Student Success System (SISSA), developed between SESu and the Federal University of Goiás (UFG), the Project "R&D: artificial intelligence to aid actions aimed at reducing dropout in higher education." The objective is to carry out research and develop experimental solutions based on Artificial Intelligence (AI) to predict academic dropout and to support pedagogical interventions in higher education. Through ICTs, the project foresees the definition of requirements, and the analysis and evaluation of the effectiveness of actions that consider the structural, organizational, and geographic scenario of the institutions. Through the Student Success System (SISSA) the aim is to integrate academic data, efficiently monitor indicators, predict student success, as well as train tutors to interact online and develop an intelligent assistant (chatbot).

4.9 *Single Access*

The Single Access Project was created to provide digital access security, modernity, transparency, user participation, integration, intelligence, monitoring, and economy. MEC reassessed the technology solutions for its electronic systems, through which opportunities for access to higher education are offered, such as the Unified

---

[5] ForPDI – Institutional development plan. Available at: https://ufcg.plataformafor.mec.gov.br/#/comunidade/forpdi?_k=axpzcg

[6] ForRisco – Institutional risk management. Available at: https://ufcg.plataformafor.mec.gov.br/#/comunidade/forrisco?_k=4gqf3l

Selection System (Sisu), the University for All Program (Prouni), and the Student Financing Fund (Fies).

4.10 Support Foundation System (SISFUNDAPOIO)

The Support Foundation System (SisFundApoio), set up by SESu/MEC and the Secretariat for Scientific Development of the Ministry of Science, Technology and Innovation (MCTI), aims to analyze requests for accreditation and authorization submitted by Support Foundations. The Support Foundations are institutions created to support research, teaching, extension, and institutional, scientific, and technological development projects, of interest to the Federal Institutions of Higher Education (IFES), and also to research institutions. According to MEC, SisFundApoio was instituted to provide greater security and speed to the analysis of Support Foundation processes carried out by representatives of the two Ministries.

5 Public Policy and Digital Transformation in Higher Education, Perspective of Promoting Educational Improvement

It is recalled that ever since the manifesto of the pioneers of the 1930s, it already boasted of an education in which the State would organize a general education plan and defended the banner of a single, public, secular, obligatory, and free school. There are more than ninety years of that manifesto prepared by renowned educators such as Fernando de Azevedo, Roldão Lopes de Barros, Anísio Teixeira, Afrânio Peixoto, Lourenço Filho, Antônio F. Almeida Junior, Roquette Pinto, Delgado de Carvalho, Hermes Lima, and Cecília Meireles, between others.

However, the National Education Plan (PNE 2014–2024) is on the verge of collapse, perhaps not so close to the failure of the previous PNE, but on the way to not achieving the proposed goals, if you consider that the PNE expresses the guidelines, goals and strategies for educational policy over a 10-year period. That is, one of the differences is that this PNE is decennial by constitutional force, which permeates regardless of the ideology of governments. Therefore, it should be a state plan!

In this context, we searched the Monitoring Panel of the National Education Plan (PNE) brings together graphs and tables of the 20 goals of the PNE 2014–2024. The data are updated every 2 years, according to the conclusion of the reports of the monitoring cycles of the goals, attribution of the National Institute of Educational Studies and Research Anísio Teixeira – Inep, according to the PNE Law (https://simec.mec.gov.br/pde/grafico_pne.php).

The PNE Monitoring Panel was released in 2016, integrating InepData, a set of BI (Business Intelligence) panels, and INEP is responsible for facilitating society"s access to the information produced. Its objective is to be an information base for educational managers of public and private networks, civil society organizations, researchers, and the press. In 2022, the PNE 2014–2024 Monitoring Panel underwent important transformations, allowing interactions between variables and enabling disaggregation by municipality.

In the aspects they express about higher education, references are contained in goal 12 of the PNE 2014–2024, access and expansion of higher education. This target includes raising the gross enrollment rate in higher education to 50% and the net rate to 33% of the population aged 18 to 24 years. Years, ensuring the quality of the offer and expansion to at least 40% of new enrollments in the public segment. From the predicted target of 50%, the current situation is at 30.3% of the target 12 A – gross enrollment rate in graduation. In indicator 12B – net enrollment rate in graduation, the expected target was 33%, and the current situation stands at 20.2%.

We are close to finishing the PNE 2014–2024, being ten years old, it needs new discussions about the achievement or not of the foreseen goals.

In this path, organized civil society also created the PNE 2014–2024 observatory (https://www.observatoriodopne.org.br/), Observatory of the National Education Plan (OPNE, 2023), coordinated by Todos Pela Educação, the initiative is made up of over twenty-eight organizations linked to Education and specialized in the different stages and teaching modalities that, together, carry out permanent monitoring of the PNE"s goals and strategies during its term. In this follow-up, it is possible to extract the results of the objectives of goal 12, proposed by the PNE 2014–2024, for higher education (Fig. 8):

The data presented by INEP and OPNE reinforce that Brazilian higher education seems, in many aspects, to be dying. A fact that, in the educational area, there is a strong link with the actions of the government and not of the State. Even though the speeches of politicians and managers demonstrate the need to improve the indicators, the follow-up reports presented by INEP and OPNE demonstrate that we are far from achieving the proposed goals of the PNE 2014–2024.

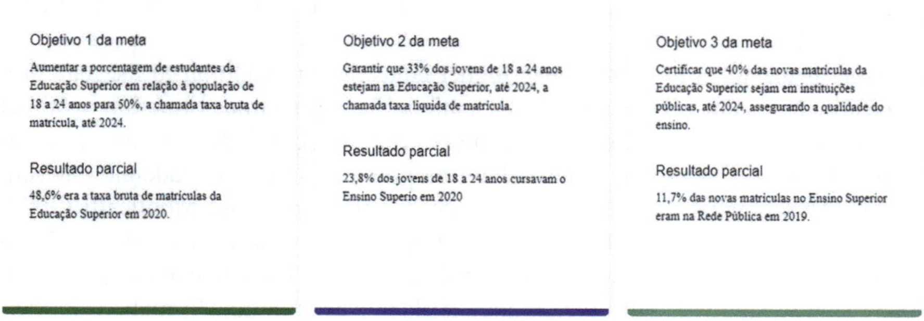

Fig. 8 Objectives of target 12 PNE 2014–2024. (*Source:* National Education Plan Observatory (OPNE). Available at: https://www.observatoriodopne.org.br/meta/educacao-superior. Accessed on 30 Jul 2023)

Another fact, not exhausting the possibilities, such as reduction of funds, contingencies, and appointment of political positions to the detriment of the technical capacity of the effective civil servants. The constant changes of technical managers, among other causes, demonstrate that the public power does not value the effective technical team (specialists, masters, and doctors) of the direct administration of the Ministry of Education (MEC).

It is clear that it is not enough to want to act in the educational area, if this alone were to solve it; we would certainly not have different graduations to act in society. A graduation would be enough for all areas of knowledge! Continued actions needed in educational policies, for a few years, to understand and manifest about aspects inherent to basic education and higher education.

No, it is not by chance, they intend to weaken the teams of trained effective servants, of the direct administration of the Ministry of Education (MEC) and its autarchies. The weakening of technical teams is a precedent for the government itself to sabotage educational programs and projects aimed at breaking the ballasts regarding the elaboration, execution, and follow-up that tend to be State policy for the maintenance of government policies.

As Santos Alves and De Oliveira Bernardi Noleto (2022) describe, we must understand such plans as results of the country"s historical, political, economic, and social moment. Yet,

> [...] it is observed, in the perspective of the implementation of the PNE and resulting policies in force in the field of higher education, especially in public universities, the indication of the expansion of vacancies in the undergraduate courses offered, however, this increase should be conditional on the quality of education. New studies are indicated that measure the quality of education offered with the expansion of vacancies, through the affirmation of the university tripod: teaching, research and extension. (Santos Alves and De Oliveira Bernardi Noleto, 2022, p. 23)

It strengthens the discourse, despite the structural and conjuncture limits of the field of Brazilian education, it is necessary to guarantee and validate the PNE, in the perspective of aiming for quality education, and education for democracy, with the objective of strengthening the construction of a democratic society. Furthermore, the democratization of higher education is only effective as a public policy when there is public funding to implement the actions provided for in the PNE (Rodrigues et al., 2021).

One of the ways to democratize higher education, in addition to the data presented in this chapter, is digital transformation. The digital transformation in higher education institutions (HEIs) arises as pressure from technological means disseminated! However, the speed is far below the possible needs of academic environments. This is what the study by Jacociunas, Verschoore, and Monticelli (2023) stresses that digital technologies and their applications have drastically declined transformed the world economy and global companies. The context also points out that the growing market dispute requires HEIs to carry out transformations in their management structures and strategies. But,

> [...] the market transformations that the Education sector in Brazil has been going through, added to a significant economic crisis, which further increased competitiveness when it

> comes to e-learning, suggest the need for a change of strategy in HEIs . In this construct, some scenarios require more urgent remodeling, such as digital transformation, which has accelerated the search for solutions for HEIs to remain economically sustainable and competitive in their market. (Jacociunas et al., 2023, p. 21)

The digital transformation in higher education is not linked to everything that the concept entails: a new way of being in teaching, a new framing of institutions, and a new way of integrating teams. However, digital transformation is not an end in itself, but rather a path to follow – a path that leads to digital maturity (Rodrigues, 2019, p. 347).

Recently, Law No. 14,533, of January 11, 2023 (BRAZIL, 2023a), instituted the National Digital Education Policy (PNED), which amends Laws No. 9,394, of December 20, 1996 (Law of Guidelines and Bases for Education Nacional), 9,448, of March 14, 1997, 10,260, of July 12, 2001, and 10,753, of October 30, 2003. This Law, in its first article, of the PNED, describes that it is structured from the articulation between programs, projects and actions of different federal entities, areas, and governmental sectors. With the aim of improving standards and increasing the results of public activities. Policies related to the Brazilian population"s access to digital resources, tools, and practices, with priority given to the most vulnerable populations. Its purpose is to integrate other federal, state, and municipal actions, projects, and programs aimed at innovation and integration in education that have technical or financial support from the federal government. In this sense, it presents the structuring axes such as I – Digital Inclusion; II – School Digital Education; III – Training and Digital Specialization; IV – Research and Development (R & D) in Information and Communication Technologies (ICTs).

For this, it will be necessary to pay attention to the smallest details if we want to achieve the digital transformation in education.

A fact that the absence and/or omission of the State leaves hostage the structures that most need support and guidance, namely, the executing units of educational policies. At this point, it is worth reaffirming Article 6, […] within the scope of the National Policy on Digital Education, the implementation of the following enabling axes will constitute the duty of the public authorities, observing the duties established in arts. 9, 10, and 11 of Law No. 9,394, of December 20, 1996 (National Education Guidelines and Bases Law) (…).

It is expensive for us, but resilience keeps us active in breaking down the barriers that, perhaps, could obscure the paths to education for all, inclusive and of quality. Being an encouragement to the improvement of public policies aimed at digital transformation in higher education.

6 Final Considerations

Exposing public policy and digital transformation in higher education is complex. In this chapter, a critical descriptive methodological approach is preferred, without intending to exhaust the theme, due to the complexity of the programs and projects of the Ministry of Education (MEC). We opted for the selection of programs and projects that stood out for their lasting permanence, in different governments, such as the Open University of Brazil (UaB) and Reuni Digital. The "program" called the Open University System of Brazil (UaB), under the management of the Coordination for the Improvement of Higher Education Personnel (CAPES) and the Digital Reunion, under the institutional responsibility of the Higher Education Secretariat (SESu), under the aegis of the Ministry of Education, to address the digital transformation in higher education in the country.

The UaB is not an HEI, considered a program formed by a consortium of federal, state, and municipal public educational institutions. Its criterion is that candidates for the vacancy must have at least finished high school and must go through a selection process, especially aimed at graduation in teaching. The most recent Reuni Digital was created with the intention of creating a Distance Federal University.

Two programs responded and met the proposal of the National Policy on Digital Education (PNED). They are structured from the articulation between programs, projects, and actions of different federal entities, areas, and government sectors. The objective is to improve standards and increase the results of public policies related to the Brazilian population"s access to digital resources, tools, and practices, with priority for the most vulnerable populations (BRAZIL, 2023a, b). Without intending to exhaust the subject due to the complexity presented in this theme, some programs and projects linked to the digital transformation in education are also sought. This study is based on the Digital Government Law (Law 14.129/2021), the Digital Government Strategy, foreseen in Decree n° 10.332/2020, Law n° 14.533, of January 11, 2023, which establishes the National Policy of Digital Education (PNED) and National Education Plan 2014–2024.

In the results of the studies, considerations presented about the digital transformation in higher education, by the supervisory bodies of the public power and organized civil society.

A highlight is the incipient maturity of the Brazilian educational system in digital government, described by the TCU in judgment 2487/2022, referring to universities, federal education institutes, and state and municipal education networks. A (Sefti), from the TCU, presented recommendations to the MEC aimed at promoting training and awareness about the digital government of education, the digital transformation of education, and digital skills for all in educational communication. In this sense, he suggested establishing a culture of digital government of education among federal entities in digital government and in the digital transformation of education. Induce actions to increase the level of maturity in the digital government of federal, state, and municipal education networks. Prepare a single and consolidated document that integrates a digital education strategy for Brazil"s plan of action, monitoring, and evaluation of the digital education strategy for Brazil. Develop a national policy on information and information technology in education, among other recommendations.

The OPNE organized civil society and the INEP autarchy of the MEC, presented data related to the fulfillment of the goals of the PNE 2014–2024. These subsidies, in our understanding, warn of a strong link between public educational policies and government actions. Reinforcing that, even with the proximity of the end of the PNE 2014–2024, the government"s actions were insufficient to achieve the goals established in the ten-year plan.

In recent years, there has been a reduction in the workforce of technical teams and low occupation of leadership positions by effective servants of the direct administration of the MEC. A situation that confirms the lack of ambition of the public power to maintain the ballast of educational programs and projects as a State policy, dedicated to the elaboration, execution, monitoring, and evaluation of the digital transformation in higher education.

For the digital transformation in higher education, in addition to UaB and Prouni Digital, other actions were presented: Connected Students Program; IFES Monitoring Portal; Educational; Digital Diploma; Icpedu – National Digital Certification Program; 360° University Platform; Combating Evasion – Platform for – Strategic Planning and Risk Management; Sissa; Single Access; Sisfundapoio.

Programs and projects are horizons; however, these policies do not always hold. Various educational programs and projects generally extinguished after the election year. Being a bad record regarding the maintenance of the achievements achieved by the previous management. Being a bad record regarding the maintenance of the achievements achieved by the previous management. It is clear that the public power did not and will not take care of the digital transformations in higher education, in the current management model that does not value the performance of the effective servants of the direct administration of the MEC. Outsourcing the attributions of the MEC, which should necessarily belong to the State, does not seem to us a minimally reasonable policy to encompass the expansion and elevation of educational quality, due to the great need for digital transformation in higher education.

Likewise, the continuity of programs and projects as a State policy, safeguarding the history of successful points or points to be improved, with a strong desire to respect public affairs that would tend to optimize public spending. It is worth remembering that State policies are strict, on the other hand, government policies are transitory and signed every election year!

In view of the elements presented in this chapter, I repeat, without intending to However, it remains to be seen which model of public policy will the digital transformation of higher education take place?. Government or State? However, a State digital transformation policy with competent professionals (education and experience) with a systemic vision would result in improvement of programs and projects aimed at the educational area. It is necessary to treat education as a State policy, in all its nuances, aiming at the elaboration, execution, monitoring, regulation, supervision, and evaluation to implement them as a policy of national results.

References

Brazil. (1961). Law No. 4024, of December 20, 1961 (LDB 4024/61). BRAZIL.

Brazil. (1988) [Constitution (1988)]. *Constitution of the Federative Republic of Brazil*. Brasília, DF: Federal Senate, 2016. 496 p. Available at: https://www2.senado.leg.br/bdsf/bitstream/handle/id/518231/CF88_Livro_EC91_2016.pdf. Accessed on: 24 July 2023.

Brazil. (1996). *Laws of guidelines and bases of national education*. LAW No. 9394, of December 20, 1996. Federal Official Gazette of December 23, 1996.

Brazil. (2005). Decree No. 5622, of December 19, 2005. Available at: http://goo.gl/YYBG2M. Accessed on: 15 Jul. 2023.

Brazil. (2006). Decree No. 5,800, of June 8, 2006. *Provides for the Open University of Brazil system – UAB*. Published in the DOU of 6.9.2006.

Brazil. (2014). [National Education Plan – PNE (2014)]. National Education Plan 2014–2024: Law No. 13,005, of June 25, 2014 – Brasília: Chamber of Deputies, Edições Câmara, 2014. 86 p. (Series legislation; n. 125).

Brazil. (2016a). Decree No. 8638, of January 15, 2016. *Establishes the Digital Governance Policy within the scope of bodies and entities of the direct federal public administration, autarchic and foundational*. Available at: https://www.planalto.gov.br/ccivil_03/_ato2015-2018/2016/decreto/d8638.htm. Accessed on: 01 Aug. 2023.

Brazil. (2016b). Decree No. 8936, of December 19, 2016. *Establishes the Digital Citizenship Platform and provides for the provision of digital public services, within the scope of bodies and entities of the direct federal, municipal and foundational public administration*. Available at: https://www.planalto.gov.br/ccivil_03/_ato2015-2018/2016/decreto/d8936.htm. Accessed on: 01 Aug. 2023.

Brazil. (2017a). Decree No. 9057, of May 25, 2017. Available at: http://goo.gl/YYBG2M. Accessed on: 15 Jul. 2023.

Brazil. (2017b). Decree No. 9094, of July 17, 2017. *Simplification of the service provided to users of public services, institutes the Individual Taxpayer Registration – CPF as a sufficient, substitute instrument for presenting citizen data in the exercise of obligations and rights, and in obtaining benefits, ratifies the waiver of notarization and of authentication in documents produced in the Country and institutes the Charter of Services to the User*. Available at: https://www.planalto.gov.br/ccivil_03/_ato2015-2018/2017/decreto/d9094.htm. Accessed on: 01 Aug. 2023.

Brazil. (2017c). Decree No. 9,235, of December 15, 2017. *Provides for the exercise of regulatory, supervisory and evaluation functions of higher education institutions and undergraduate and graduate higher education courses in the federal education system*. Available at: https://www.planalto.gov.br/ccivil_03/_ato2015-2018/2017/decreto/D9235.htm. Accessed on: 01 Aug. 2023.

Brazil. (2019a). Decree no. 9723, of March 11, 2019. *Establish the Individual Taxpayer Registration – CPF as a sufficient instrument and substitute for the presentation of other documents of the citizen in the exercise of obligations and rights or in obtaining benefits and regulating provisions of Law No. 13,460, of June 26, 2017*. Available at: https://www.planalto.gov.br/ccivil_03/_ato2019-2022/2019/decreto/d9723.htm. Accessed on: 01 Aug. 2023.

Brazil. (2019b). Decree No. 9756, of April 11, 2019. *Establishes the single portal "gov.br" and provides for the unification rules of the federal government's digital channels*. Available at: https://www.planalto.gov.br/ccivil_03/_ato2019-2022/2019/decreto/d9756.htm. Accessed on: 01 Aug. 2023.

Brazil. (2020a) Ordinance No. 343, of March 17, 2020. *Provides for the replacement of face-to-face classes with classes in digital media while the pandemic situation of the new coronavirus – COVID-19 lasts*. Available at: Accessed https://www.in.gov.br/en/web/dou/-/portaria-n-343-de-17-de-marco-de-2020-248564376. Accessed on: 24 Jul. 2023.

Brazil. (2020b). Ordinance No. 433, of October 22, 2020. *Establishes the strategic guidance Committee – COE for the elaboration of initiatives to promote the expansion of higher educa-*

tion through digital means in Federal Universities. Available at: (https://www.in.gov.br/web/dou/-/portaria-n-433-de-22-de-outubro-de-2020-284699498). Accessed on: 29 Jul. 2023.

Brazil. (2020c). Ordinance No. 434, of October 22, 2020. *Establishes a working group with the purpose of subsidizing the discussion, elaboration and presentation of strategies for expanding the supply of higher education courses, in the distance education modality – EaD, in federal universities*. Available at: (https://www.in.gov.br/en/web/dou/-/portaria-n-434-de-22-de-outubro-de-2020-284699573). Accessed on: 29 Jul. 2023.

Brazil. (2021a). [*National Institute of Educational Studies and Research Anísio Teixeira* (2021)] (Inep). Higher Education Census 2021: Statistical notes. Brasília, DF: Inep.

Brazil. (2021b). Law No. 14,129, of March 29, 2021. *Provides for principles, rules and instruments for Digital Government and for increasing public efficiency*. Available at: https://www.in.gov.br/en/web/dou/-/lei-n-14.129-de-29-de-marco-de-2021-311282132. Accessed on: 27 Jul. 2023.

Brazil. (2022). Federal Court of Accounts Judgment 2487/2022 – PLENARY, TC 043.945/2021-0. *First Chamber. Rapporteur: Minister Benjamin Zymler*. Process TC 043.945/2021-0. MINUTES 42/2022 – Plenary Session. Monitoring Report (RACOM). Brasília, DF, Session 11/01/2022.

Brazil. (2023a). Law No. 14,533, of January 11, 2023. https://www.planalto.gov.br/ccivil_03/_Ato2023-2026/2023/Lei/L14533.htm. Accessed on: 24 Jul. 2023.

Brazil. (2023b). [Ministry of Education (2021)]. *Portal of the Ministry of Education*. Available (https://www.gov.br/mec/pt-br). Accessed on 03 Jul. 2023.

Amaro, R., & Baxto, W. (2013). *Darcy Ribeiro's Pedagogical Thought: from the necessary university to the proposal for the creation of the Open University in Brazil*. In: 19th International Congress on Distance Education, 2013, Salvador. 19th International Conference on Distance Education.

Brazilian Internet Steering Committee (CGI.br). (2023). *Cultural equipment expands its online presence, but the Internet offer for the public is still limited: Methodological report*. São Paulo: CGI.br. Available at: (https://cetic.br/pt/noticia/equipamentos-culturalis-ampliam-sua-presenca-online-mas-oferta-de-internet-para-o-publico-e-ainda-reçada/).

Brazilian Institute of Geography and Statistics (IBGE). (2022). *Brazilian Census* (p. 2022). IBGE.

Jacociunas, T., Verschoore, J. R., & Monticelli, J. M. (2023). Digital transformation of higher education institutions: A framework for strategic decision-making. *International Journal of Higher Education, 10*, e024036. https://doi.org/10.20396/riesup.v10i00.8665499

Mundim, K. C. (2006). *(ORGs). Distance learning in Brazil: Problems and challenges. Challenges of distance education in teacher training – Brasília* (pp. 119–126). Secretariat for Distance Education.

Observatory of the National Education Plan [OPNE]. (2023). Available at: https://www.observatoriodopne.org.br/meta/educacao-superior. Accessed on 30 Jul 2023.

Rodrigues, A. A. B. (2019). The digital transformation in higher education: Obligation? Option? Opportunity? *FORGES Lecture, 9*, 1–388.

Rodrigues, D. P., Freitas, S. T., & Pezavento, K. (2021). National education plans (2001–2010 and 2014–2024) in the field of higher education: Advances and/or setbacks. *International Journal of Higher Education, 8*(00), e022005. https://doi.org/10.20396/riesup.v8i00.8660972

Santos Alves, H., & De Oliveira Bernardi Noleto, S. (2022). National Education Plan in Brazil: Mechanism for the democratization of access to higher education. *Revista Electronica Actualidades Investigativas En Educación, 22*(2), 1–27.

Silva, W. B. (2007). *Public policies in distance education: From legal to real* (p. 75). Monograph (Specialization in Distance Education). University of Brasília.

Unesco. (2015). *Unesco Science Report, Towards 2030, Overview and Brazilian Scenario*. Available at: https://unesdoc.unesco.org/ark:/48223/pf0000235407_eng. Accessed on 01 Jul 2023.

Augmenting Open and Distance Learning Through Gamification

S. K. Pulist and Ramesh Chander Sharma

1 Introduction

Distance learning and online learning have proved their impact at various levels of education during the past decades. The comfort of the students and teachers in using educational resources at a distance has made educational institutions rethink the integration of technology in their instructional design at different levels. The enabling provisions by the University Grants Commission of India through the promulgation of the University Grants Commission (Open and Distance Learning Programmes and Online Programmes) Regulations, 2020, have helped the Open and Distance Learning (ODL) institutions to make use of online resources to the extent of 40% of the course content (UGC, 2020). This has made the ODL institutions do experiments with the instructional design of their program by integrating different learning elements to enhance the efficacy of their instructional methodology. Nowadays, gamification is being tried in other areas such as production, marketing, sales, health, staff training and development, entertainment, sports, finance, social, ecological settings, etc., including education (Miciuła & Miciuła, 2015). Gamification in the educational context is not very old, though it is gaining momentum daily. With continued growth, people realize the importance of introducing game mechanics in education.

Games have been part of every culture and have been primarily used as a tool for entertainment. The advent of the Internet has helped the gaming process to be online. Resultantly, a vast gaming industry has come up. The concept of

S. K. Pulist (✉)
Indira Gandhi National Open University, New Delhi, India
e-mail: skpulist@ignou.ac.in

R. C. Sharma
School of Global Affairs, Dr. B R Ambedkar University Delhi, New Delhi, India

A. de Bem Machado et al. (eds.), *Digital Transformation in Higher Education Institutions*, EAI/Springer Innovations in Communication and Computing,
https://doi.org/10.1007/978-3-031-52296-3_5

gamification is comparatively new, but the tools and techniques it uses are not new. It is believed that Nick Pelling first used this concept in 2002 (Zichermann & Cunningham, 2011), though it became well-known only after the availability of the Internet in 2010. Gamification has become a growing trend in the twenty-first century. Organizations have been using gaming elements in the professional development of human resources. The concept is liked more by the Y and Z generations, people who are part of the workforce (Miciula & Miluniec, 2019).

In the sales, sports, insurance, and leisure industries, customers are motivated to earn more and more points as bonuses, which are discounted on purchasing some valuables. Organizations can use gamification to enhance the productivity of their employees and the quality of their products (Para, 2021). Challenges are posed before the users in the form of games, which they have to overcome level by level and earn bonus points. In education and professional training, the techniques of the game are used to solve real-life problems, making the process more exciting and engaging. The core components of gamification are the participant's engagement, change, reward behavior, and problem-solving (Tkaczyk, 2012). According to Miciula and Miluniec (2019), the activities performed by the participants are directed towards specific goals other than mere games or entertainment. With the help of the introduction of gaming elements, the boring stuff can be made interesting, and thus, participants can be motivated to take an active part in the process. The participants derive motivation through pleasure, which is attained by "overcoming the next achievable challenges, competition, cooperation, etc." (Miciula & Miluniec, 2019).

While online learning is already using different components of synchronous and asynchronous modes for making learning more interactive, outcome-oriented, and enjoyable, conventional distance learning also can be revolutionized with the implementation of game mechanics in instruction delivery. Gamification provides a learner-centered environment and, thus, becomes more engaging and promotes learning through problem-solving by posing challenges gradually level by level. It can be an enabling and game-changer tool for enriching the learning experiences of distance learners. The ODL institutions can use gamification as a teaching–learning strategy along with other strategies, keeping in view the objectives of the content to be delivered. It can be implemented as an essential tool as part of the instructional design.

2 Context

The concept of gamification is one of the critical contributions of Education.4.0. To meet the human resource and workforce requirements for Industry 4.0 in the future; the education system needs to find innovative ways to impart education before it becomes redundant, more so when generations Y and Z are in focus. The education system needs to develop a workforce for the jobs that are yet to be created or conceptualized. The open and distance learning system is seen as a scalable system that can cater to the massive demand of such a specialized human resource. The gamification framework provides the potential instructional design that can meet all the challenges in the process. Coupled with an open and distance education learning system, the gamification design can deliver the desired results. The present study explores the potential of the implementation of gamification in open and distance learning to cater to mass-level human resource requirements. The study has adopted a descriptive method followed by a review of published articles and documents on gamification.

3 Gamification in Education

Gamification in education is presenting the course content before the students through gaming elements in a game-equivalent context to motivate and engage them with the content (Pappas, 2013). Hanghoj (2013) defined gamification as the "use of game elements in non-entertainment contexts to promote learning." Gamification is not a simple game. Also, it is not a process to replace teaching with gaming. Instead, it is a technique of using game elements and mechanisms to make the learning activities more encouraging, engaging, compelling, and enjoyable. Focusing on teaching content and designing how it is presented rather than the technology itself is at the core of gamification.

The concept of gaming involves the accumulation of bonus points, badges, and awards to motivate the participants to get actively involved in the gaming process. Competition and collaboration are essential components of game mechanics. Shahri et al. (2019) consider social recognition as one of the positive reinforcement and awards. These components are added to the process, considering the type of behavior to be encouraged in the participants. Research shows that the same reward becomes dull and loses its sanctity and usefulness with time. Therefore, challenge and reward need to be more prominent in the passage, which is at the core of the gaming concept. Here also, the points/reward increase with the level of difficulty, which fetches a more satisfying experience for the participant when accomplished. For them to keep performing on the gaming platform, the gaming elements need to be arousing intrinsic motivation in the participants.

Gamification uses "game-based mechanics, aesthetics and game thinking to engage people, motivate action, promote learning and solve problems" (McGrath & Bayerlein, 2013). Simões et al. (2013) distinguish the mechanisms of gamification drawn from games as feedback, rewards, competition and opportunities for interaction, well-defined goals, and levels of commitment leading to the acquisition of a new skill set. The intensity of challenges increases with every level, building the confidence and commitment of the participants and giving a sense of development and achievement.

The unique feature of gamification pedagogy is that the games are used in a pedagogical context for augmenting learning. The researchers' response on the effectiveness of gamification in teaching and learning has been mixed. While some have emphasized that gamification helps learners clearly understand different concepts, the relationship between two or more concepts, and the applicability of that knowledge in a specific context (Subhash & Cudney, 2018; Limantara et al., 2019; Bai et al., 2020; Hamari et al., 2014), others have observed that merely applying gaming components in education will not fetch the desired results. Instead, they emphasized the need for good pedagogical design and contextual considerations for using gamification in the teaching and learning process (Hamari et al., 2014; Dichev & Dicheva, 2017).

The introduction of gamification in the pedagogical process is proposed because the students take more interest in game-based learning activities and try to master the situation (Jadel et al., 2021). The elements and mechanics used in gamification naturally appeal to the younger generation and thus strongly impact their lifestyle, including how they learn new things. The games to be used as teaching and learning devices should be carefully considered, keeping in view the intended impact of each of the games on the performance of the students (Dichev & Dicheva, 2017).

4 Gamification Pedagogy

Nieto-Escamez and Roldán-Tapia (2021) consider quizzes as simple versions of gaming techniques used in pedagogical processes invariably. They allow the students to test their knowledge instantly and receive feedback. The significant strategies employed in the process are the challenge – the student is required to overcome the difficulties (Koivisto & Hamari, 2019); the immersion – an attempt to immerse the participant into a narrative presented by audio-visual integration (Concannon et al., 2019); and social-based gamification – encouraging the participant to develop strategies to compete and collaborate in the environment (Romero, 2017). The course-based games are designed in such a way that they introduce new concepts progressively. This helps the students to master the concept and further use it at the next level, synthesizing it with their prior knowledge. This way, the students can develop new strategies and problem-solving techniques level after level in a linear way (Varonis & Varonis, 2015).

Motivation is claimed to be one of the most important predictors of students' learning outcomes since it significantly impacts the time and effort spent engaging with the curriculum for learning. The gamification activities are reported to enhance the intrinsic and extrinsic motivation of the students in that they address the basic psychological needs of autonomy, relatedness, and competence (Richter et al., 2015). Landers (2014) argues that the student's performance, commitment, the feedback received by them, the complexity level of the activity, and constraints posed by the situation are essential drivers of student motivation. In another parlance, Huang and Hew (2018) argue that gamification should contain specific, understandable goals, instant feedback, performance indicators, equilibrium among the student's skills, the intensity of challenge and perceived importance, and the value of the activity to motivate the students to accomplish the game. The educationists need to effectively use the gaming elements to cater to the students' differing learning styles in the educational contexts.

Enactment and role-play are the crucial components of gaming elements. Crossword exercise is another example of a simple game element used in the gaming environment. Other gaming elements identified by Nieto-Escamez and Roldán-Tapia (2021) through a review of different studies are, viz., competition, cooperation, challenge, points, leader board, videoconference, escape room-based procedure, progress bar, avatars, challenges, levels, progress, missions treasure hunt, photo quiz, Imaging quiz, notebook quiz, gamebook, pictures, videos, serial cues, simulation, and dice.

5 Gamification Design

The introduction of gamification in education satisfies the requirements of flow theory, which propagates that enjoyment and engagement are maximized at an optimal psychological and physical state. The game mechanics ensure that the attention and focus of the students are captured thoroughly so that their enjoyment and engagement are enhanced (Huang & Hew, 2018). However, in order to develop an environment that caters to different factors such as the personality of the students, content needs and student's needs, value system, and level of motivation of the students, the developer of gaming elements should understand their mechanics and dynamics appropriately (Bennani et al., 2021).

The gamification design primarily focuses on changing the individual's behavior. This behavior modification motivates the individual to participate in the learning activity disguised as the gaming activity. The element of pleasure arouses intrinsic motivation in the student, who thus seeks full involvement in the activity. The sense of impending victory converts the activity into a mission to be accomplished without being bothered about the failure of attempts (Para, 2021). A similar context can be replicated in education to augment learning. Para (2021) discussed different aspects of gamification design to enhance the effectiveness of the gaming elements in engaging the students and boosting their motivation, which is a challenge in the distance learning environment. According to Deterding et al. (2011), the goal of gamification design is to inspire people to change their behavior and enhance their involvement in the task at hand. It promotes game-thinking techniques to solve the problems (Tkaczyk, 2010). Judicious use of gaming elements can expand the scope of their use in different contexts. The tokens, badges, and stickers the users earn can be converted into something valuable to him/her.

Para (2021) emphasizes that the reward and punishment system is essential to motivate the participants to participate actively in the gamified activity. The set procedure of the game may include earning points, ranking based on scores, and presenting results. From the systemic point of view, these staged achievements help the teachers to track the progress of the participants. In education, the gamified elements can fetch points to the students, which can be reckoned toward the final class grades of the participants. Any activity undertaken by the students in the class can be mechanized as a game, including presentations, essay writing, quiz attempts, or mini-projects linked to earning scores/scores.

Iverson and Colky (2004) argued that the scenario-based pedagogical design best suits game-based learning. The scenario-based design focuses on the delivery of content in an asynchronous mode, which enables the learners to interact with the content independently. A goal-based scenario is created in the learning environment based on the objectives of the content to be transacted with the students. The learning outcomes of the content are decided before applying the gamification design. The students interact with the gaming elements and pursue the objective by "practicing key skills and using content knowledge" (Iverson & Colky, 2004). The students face realistic scenarios while performing a critical role they may have to perform in real life. The supporting material needs to be provided to the students with academic support from the teacher whenever needed by the students. These gaming techniques in instructional design focus on work-based challenges and problems from authentic scenarios that make it easy for the students to transfer their learning to real-world contexts. The game-based interaction facilitates the teacher to track the progress of the students and chalk out interventional strategies to help them master the concept.

6 Gamification Design Framework

Hitchens and Tulloch (2018) report that the younger generation has grown amid video games and social networks, and something of this sort would be more suited to their learning styles. Therefore, they justify using gamification as an innovative and alternative tool for teaching and learning, thereby exploiting the habits of young people and how they learn. The specific "media consumption patterns" of the younger generation have made them fundamentally different from the earlier generations (Bourgonjon et al., 2010). Bourgonjon et al. (2009) have made similar observations about the young generation of students who have acquired specific technical skills and thinking styles and have different learning preferences. Thus, they need new pedagogical approaches for their engagement in learning.

The concept of gamification may not be implementable in all the disciplines. There may be some disciplines that are not suitable for gamification. Therefore, the concept should be carefully integrated into the selected course content (Antonaci et al., 2019). Distance education is yet to be considered as a new area for the implementation of gamification, requiring empirical evidence. Freitas and Silva (2020) argued for developing a theoretical framework for integrating games with the course content and relating it to the specific learning outcomes. Herzig et al. (2015) also argued that the introduction of gamification for learning needs a specific instructional design. The researchers proposed a four-level framework for applying game mechanics in learning as "business modeling and requirements, design, implementation, and monitoring and adaptation." A team of stakeholders such as IT experts,

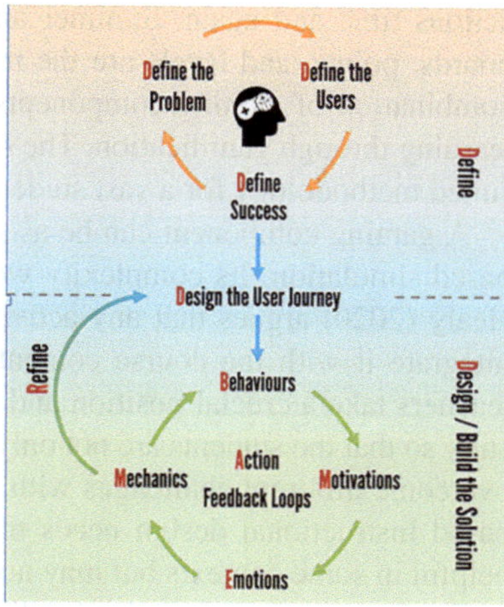

Fig. 1 Gamification design framework. (Source: KSU (2022). Pedagogy for online teaching: Gamification)

domain experts, gamification experts, business experts, and end-users should be involved to develop an effective instructional design. Poor collaborative approach, lack of commitment on the part of the stakeholders, poor design, and inability to manage the process may lead to complete failure of implementing the game mechanism in learning (Shahri et al., 2019). KSU (2022) proposed a comprehensive design framework (define, design, and refine) to introduce gamification in teaching and learning, as presented in Fig. 1.

The gamification environment can be customized to give a personalized learning experience to the student. It can accommodate teacher-directed/ self-paced learning while facilitating interaction with the learning content (Wood et al., 2010). It can facilitate the application of theoretical knowledge to real-world experiences and thus bridge the gap between the theory and acquired skills. Setting learning outcomes for a topic, designing a framework to achieve the outcomes, deciding the mechanism to measure the progress, identifying the participants and their needs, and building a community of participants of the game would be crucial to make the gamification successful.

7 Gamification Tools

Huizinga (1944) describes that apart from the real world, a game is an activity voluntarily in its own time and place according to set rules. The gaming elements deploy different mechanisms to make them engaging for the users. Different elements such as "story, tasks, points, levels, challenges and missions, badges, rankings, reward system" are used as mechanics to promote games and motivate the users to take on new challenges. Rewards are essential in gamification since they are the natural motivating agents. They lead to reinforcement of rewarding actions and thus influence the behavior, which further motivates them to repeat the positive actions time and again. Summer and Aydin (2018) reported that badges, leaderboards, points, and levels are the most used gaming elements in ODL. The right combination of gaming components is crucial for sustaining students' interest in learning through gamification. The researchers argued that gamification is the most suited methodology for a vast student base, as in the ODL system.

A gaming component can be as simple as a quiz and as complex as a scenario-based simulation. Its complexity would depend on the topic for which it is used. Healy (2020) argues that any activity can be gamified, though it requires skill to integrate it with the course content and learning objectives. In this process, the teachers take a crucial position and need to manage the gaming activities strategically so that the students are not only willing to participate in it but also are ready to overcome different challenges with multiple levels of difficulty. The gamification-based instructional design needs to be implemented judiciously since it may be helpful in some contexts but may not be as effective in other contexts.

Some examples of gaming applications are KnowRe – a math learning application customized to adapt to the traits of the individual students, Duolingo – an application that offers over 68 languages that can be learned by the students at their own pace; ClassDojo – a behavior improvement application based on badges, levels and leaderboards, McDonald's – training application for improving customer service, LIFESAVER – a simulator to provide training for cardiopulmonary resuscitation (CPR), and, Deloitte (Will You Fit into Deloitte) – is used to assess the suitability of the recruits for the company (Engine, n.d.). Free Rice is another gaming website that helps students to check their acquired knowledge. The Trivia – a gaming platform developed by Price Waterhouse Coopers, was used for training over 3000 interns during the Pandemic time (Gedetsis, 2021). Microsoft created Ribbon Hero 2 gaming application for training people in different utilities of the MS Office suite (Fallows, 2011). Some other gaming tools used in education are Credly, Openbadges, Kahoot, Brainscape, Second Life, Minecraft, and TopHat (UoW, n.d.).

8 Open and Distance Learning and Gamification

In distance learning, gamification has an added advantage since the platform can be linked with hints and explanations of specific concepts and problem areas, real-time feedback, and tracking of the activities and performance of the students while in action. The students are deeply involved with the environment cognitively, affectively, and behaviorally, and create a constructive learning experience for them. The use of gaming techniques in distance education can improve the performance of the learners in different subjects (Escamilla & Martinez, 2020). When used even for boring topics, it can improve student participation due to the unique opportunities it provides. It allows the students to interact with the content and apply their acquired knowledge on a specific topic in an entertaining, competitive, collaborative, and time-smart way.

The educational games encourage the players to apply their analytical and problem-solving skills in-depth, forgetting the orientation of time, surroundings, and the prevailing environment (Ozelik et al., 2013). With the attainment of each level, the participants acquire new skills and design new strategies to address the challenges faced in the next level. Ozelik et al. (2013) reported that gamification forces the students to use prior knowledge, contextualize the knowledge to new situations, correctly apply that knowledge, and keep on learning from the reflections in order to move to the next level in a game. Karaolis (2021) reports active learning, continuous learning, improved productivity, enjoyable experiences, and performance tracking as the potential advantages of using gaming mechanics in distance learning.

9 Effectiveness of Gamification in ODL

Most empirical studies have found that gamification has enhanced student engagement and performance, thereby discouraging the drop-out rate to a large extent (Khalil et al., 2018; Rohan et al., 2020). Escamilla and Martinez (2020) revealed that using gaming tools in distance education enhanced understanding by 85% in the case of complex topics, and a lesser failure rate was seen among the students. By interacting with learning games, the students can learn new cognitive processes, investigate a topic, and acquire abilities and strategies to treat the acquired information in a new way.

Chapman and Rich (2018) claimed that using gaming elements in a course motivated the students to participate more than in a traditional course. The age and gender, demographic category, and status did not have any impact on the motivation of these students. Ferianda et al. (2018) used course content, discussion forums, assignments, and quizzes as gaming components and found that they had a positive impact on the performance of distance learners. The instant feedback enhanced their motivation to be involved with the activity longer.

Wankel et al. (2010) reported that game-based learning is more effective than the traditional methods of learning since they facilitate four times greater knowledge transfer and ten times greater knowledge retention. In another study, Vaibhav and Gupta (2014) reported that using game elements in instructional design enhances course completion and pass-out rates. The study by Smiderle et al. (2020) found that the students who interacted with the gaming environment had higher average scores than the participants who did not participate in the said environment. The researchers noted a significant improvement in the behavior of the experimental group. Demet et al. (2022), through their study, revealed that using animation as a gamification component enhanced the motivation and knowledge of the distance learners of the nursing program. They had improved attention and learning satisfaction after using the animation.

However, the study conducted by Jadela et al. (2021) reports no significant impact of gamification design on the academic performance and drop-out rate in open and distance learning program. The researchers emphasized that the phenomenon of implementation of gamification in distance learning needs to be studied in-depth since it takes more than the gamification design to make the framework effective. The gamification design needs to focus on how the students would interact with the game elements (Khalil et al., 2018; Shahri et al., 2019). Deterding (2015) argued that the effect of gamification depends on the extent of autonomy given to the student, the level of competence of the student, and the relatedness of the game element to the course content. The extent of engagement of the game element with the course content would play an essential role in deciding its effectiveness (Renz et al., 2016).

10 Advantages of Gamification

The broad objectives that the gamification techniques are expected to meet are: to engage students, enhance motivation, optimize learning, lead behavioral changes, promote socialization, and attach a purpose to learning (Dichev & Dicheva, 2017). The researchers also claimed that gamification in education is assumed to be capable of improving attendance and active participation in student-content interaction. Perhaps these merits of gamification have made instructional designers employ gaming techniques in different educational contexts and levels (Koivisto & Hamari, 2019). The use of gaming techniques in distance education can bring the following advantages to the teaching and learning process:

11 Active Learning

In open and distance learning, the gaming elements are used for teaching in such a way that they force the students to become active and interact with the content at a deeper cognitive level since they would be required to use that knowledge for the next level. Suppose they become a passive recipient of information at the primary level. In that case, they will not be able to successfully cross the desired levels to accomplish the learning objectives of the game.

12 Adaptive to Learning Needs

Distance learners usually come from different backgrounds impacted by age, gender, prior education, socio-economic status, geographical location, etc. One size cannot fill the learning styles of all learners. Therefore, the teaching tools must appreciate and accommodate this heterogeneity of the students. The adaptive tools can meet these objectives of instructional design. The gamification, due to its very characteristic of adaptability, is more suited to the styles and requirements of distance learners. The gaming interventions are scalable. Each level may serve the purpose of accommodating the self-paced learning for a heterogeneous student cohort.

13 Regular Learning

The mechanism of the gaming environment, specific learning outcomes of the content, instant support, performance monitoring, and collaborative and competitive edge enhance the intrinsic and extrinsic motivation of the students. Sometimes, the students compete with themselves and try to beat earlier scores in less time. The balance between the level of challenge and skills acquired by the students helps them reach an optimal state of experience and ecstasy, as propounded by Csikszentmihalyi (1990) under the Flow Theory. The visibility of progress indicators further boosts the morale of the students.

14 Enjoyable Experience

The gaming elements, by default, ensure that a fun element is attached to the learning process. This makes the students engage with the game for a longer duration. Learning with fun motivates the students to engage more and more with the game, cross the levels individually, and experience enjoyment as part of the learning process.

15 Performance Monitoring

Nieto-Escamez and Roldán-Tapia (2021) argue that the game mechanics allow tracking and monitoring of students' performance, learning outcomes, and behavior in the learning environment in a non-invasive manner. Gamification facilitates the measurement of results in light of the proposed objectives. The student's interaction is monitored and tracked extensively in a gaming environment. The moves of the students can be recorded at every click. Data collection at a subtle level enables the system to give instant feedback. It hints at the right direction and provides assistive resources to enable the student to complete the performance levels successfully. All this can happen by focusing on the learning gap between the specific learning outcomes expected and their actual performance on the task. The process builds a scaffolding support environment around the students and ensures their total commitment, leading to improved academic performance. Consequently, the students can cover more content in less time.

16 Instant Feedback

Feedback is an enabling tool for augmenting the teaching and learning process. It is equally vital in gamification design also. The participants could be informed about their scores and the scores of other participants to ascertain their ranking among them. This could be presented as a score table (Para, 2021). However, the rules of the games should be pre-defined so that there is no scope for ambiguity in their interpretation among the participants.

17 Challenges

The significant barriers to the popularization of gamification in the teaching and learning process, as narrated by An et al. (2020), are lack of time with teachers, knowledge to integrate games with teaching and learning activities, financial resources, availability of compatible games with the course content, and perception of students about gamification.

Technology versions: The teachers, developers, and the students are required to work with the available technology. All the teachers and students do not have the state-of-the-art devices with them. Likely, the software uploaded on the device may be unable to play/run the games developed on the latest software versions (Nieto-Escamez & Roldán-Tapia, 2021). Choosing and creating gaming components that are relevant, useful, and best suited to the content and topic to be covered is another major challenge.

Designing effective strategies using game mechanisms requires expert teamwork, which is a cost-intensive and time-consuming affair. A team of stakeholders, such as IT experts, domain experts, gamification experts, business experts, and end-users, must be involved in the design and development process of the educational games. Organization of human resources at this level may create a financial constraint for open and distance learning institutions. However, once designed, developed, and implemented, it may be conducive since the strategy can cater to a large segment of students simultaneously. In the absence of adoption of a collaborative approach, lack of commitment on the part of the stakeholders, poor gamification design, and inability to manage the process may lead to complete failure of the process of implementation of the game mechanism in learning. Injudicious use of multiple strategies, though not required by the content, may make the gaming structure complex and disproportionate to the level of challenge required by the topic, which may prove counterproductive.

18 Implications

The chapter presents a review of the implementation of the concept of gamification in different areas, including open and distance learning. It explores the potential usage of game design and game elements in the teaching and learning process. The analysis will be helpful to the instructional designers and other stakeholders who were not able to exploit the potential of the gamification framework.

19 Conclusions

The continuous development of Information Communication Technology (ICT) facilitates educational institutions and educationists to try innovative ways to engage students and adopt the student-friendly instructional design. Gaming techniques can augment the learning experiences of the students in a big way. Therefore, instructional designers must integrate gamification into their pedagogical processes and develop techniques for delivering effective teaching. This would take advantage of the inclination of the younger generation to use social media, online video games, and other simulation programs. Thus, the use of game mechanics in instructional design in the ODL system will suit the technological bent of mind of the young students. This will make the knowledge acquisition a playful activity rather than a severe learning process for the students. According to McGrath and Bayerlein (2013), the students will experience optimal flow in their interaction with the content and acquire critical thinking and problem-solving skills by interacting with the game-based learning environment.

20 Recommendations

The systematic approach in designing gaming elements will help prevent unexpected losses and minimize any risk/side effects of implementing the gamification design. The personal traits of the students are likely to impact the effectiveness of gamification on their learning and knowledge acquisition. Therefore, there is a need to thoroughly study the impact of these factors and customize the gaming strategies in tune with the traits of the students. While designing instruction through gamification, the crucial points such as learning objectives, pedagogical environment, student profile, feasibility, cost efficiency, and measurability of outcomes should guide the process.

References

An, Y., Zhu, M., Bonk, C. J., & Lin, L. (2020). Exploring instructors' perspectives, practices, and perceived support needs and barriers related to the gamification of MOOCs. *Journal of Computing in Higher Education*. https://doi.org/10.1007/s12528-020-09256-w

Antonaci, A., Klemke, R., & Specht, M. (2019). The effects of gamification in online learning environments: A systematic literature review. *Informatics, 6*(3), 1–22. https://doi.org/10.3390/informatics6030032

Bai, S., Hew, K. F., & Huang, B. (2020). Does gamification improve student learning outcome? Evidence from a meta-analysis and synthesis of qualitative data in educational contexts. *Educational Research Review*. https://doi.org/10.1016/j.edurev.2020.100322

Bennani, S., Maalel, A., & Ghazala, H. B. (2021). Adaptive gamification in eLearning: A literature review and future challenges. *Computer Applications in Engineering Education, 30*, 628–642. https://doi.org/10.1002/cae.22477

Bourgonjon, J., Valcke, M., Soetaert, R., & Schellens, T. (2010). Students' perceptions about the use of video games in the classroom. *Computer and Education, An International Journal, 54*, 1145–1156. Retrieved from http://www.sciencedirect.com/science/article/pii/S0360131509003121

Bourgonjon, J., Valcke, M., Soetaert, R., & Schellens, T. (2009). Students' perceptions about the use of video games in the classroom. *Computers & Education, 54*(4), 1145–1156. https://doi.org/10.1016/j.compedu.2009.10.022

Chapman, J. R., & Rich, P. J. (2018). Does educational gamification improve students' motivation? If so which game elements work best? *Journal of Education for Business, 93*(7). https://doi.org/10.1080/08832323.2018.1490687

Concannon, B. J., Esmail, S., & Roberts, M. R. (2019). Head-mounted display virtual reality in post-secondary education and skill training. *Frontiers in Education, 4*, 80. https://doi.org/10.3389/feduc.2019.00080

Csikszentmihalyi, M. (1990). *Flow: The psychology of optimal experience*. Harper and Row.

de Freitas, M. J., & da Silva, M. M. (2020). Systematic literature review about gamification in MOOCs. *Open Learning: The Journal of Open, Distance and e-Learning*. https://doi.org/10.1080/02680513.2020.1798221

Demet, I., Berna, D., & Ayse, K. (2022). Effectiveness of the use of animation and gamification in online distance education during pandemic. *Computers, Informatics Nursing*. https://doi.org/10.1097/CIN.0000000000000902

Deterding, S. (2015). The lens of intrinsic skill atoms: A method for gameful design. *Human-Computer Interaction, 30*(3–4), 294–335. https://doi.org/10.1080/07370024.2014.993471

Deterding, S., Dixon, D., Khaled, R., & iNacke, L. (2011). From game design elements to gamefulness: defining gamification. In *Proceedings of the 15th international academic mind trek conference: Envisioning future media environments* (pp. 9–15). https://doi.org/10.1145/2181037.2181040

Dichev, C., & Dicheva, D. (2017). Gamifying education: What is known, what is believed and what remains uncertain: A critical review. *International Journal of Educational Technology in Higher Education, 14*(9), 1–36. https://doi.org/10.1186/s41239-017-0042-5

Engine, E. (n.d.). *Get your game on: Benefits of gamification in online learning [Blog]*. Retrieved April 13, 2022 from https://blog.extensionengine.com/

Escamilla, R. D., & Martinez, B. E. (2020). Gamification as a basic tool in distance education. *Journal of Computer Technology, 4*(12), 17–23. https://doi.org/10.35429/JCT.2020.12.4.17.23

Fallows, J. (2011, April 28). The return of Clippy. *The Atlantic*. Retrieved April 13, 2022 from https://www.theatlantic.com/technology/archive/2011/04/the-return-of-clippy/238032/

Ferianda, M. R., Herdiani, A., & Sardi, I. L. (2018). Increasing students interaction in Distance Education using gamification. In: *6th International Conference on Information and Communication Technology (ICoICT)*, IEEE. 3–5 May 2018. https://doi.org/10.1109/ICoICT.2018.8528765

Gedetsis, J. (2021, April 7). *Why the future of eLearning might be gamification*. Retrieved April 14, 2022 from https://www.nutanix.com/theforecastbynutanix/industry/why-the-future-of-distance-learning-might-be-gamification

Hamari, J., Koivisto, J., & Sarsa, H. (2014). Does gamification work? - a literature review of empirical studies on gamification. In *Proceedings of the annual Hawaii international conference on system sciences*, pp. 3025–3034, https://doi.org/10.1109/HICSS.2014.377

Hanghøj, T. (2013). Game-based teaching: Practices, roles, and pedagogies. In new pedagogical approaches. In S. de Freitas, M. Ott, M. M. Popescu, & I. Stanescu (Eds.), *New pedagogical approaches in game enhanced learning: Curriculum integration* (pp. 81–101). IGI Global. https://doi.org/10.4018/978-1-4666-3950-8

Healy, M. (2020). *Points on the board: How I gamified my PhD. by publication progress*. Retrieved 5 April 2022 from https://mojohealy.com/post/points_on_the_board/

Herzig, P., Ameling, M., Wolf, B., & Schill, A. (2015). Implementing Gamification: Requirements and Gamification platforms. In *Gamification in education and business* (pp. 431–450). Springer. https://doi.org/10.1007/978-3-319-10208-5_22

Hitchens, M., & Tulloch, R. (2018). A gamification design for the classroom. *Interactive Technology and Smart Education, 15*(1), 28–45. https://doi.org/10.1108/ITSE-05-2017-0028

Huang, B., & Hew, K. (2018). Implementing a theory-driven gamification model in higher education flipped courses: Effects on out-of-class activity completion and quality of artifacts. *Computers & Education, 125*, 254–272. https://doi.org/10.1016/j.compedu.2018.06.018

Huizinga, J. (1944). *Homo Ludens*. Routledge.

Iverson, K., & Colky, D. (2004). Scenario-based e-learning design. *Performance Improvement, 43*(1), 1–7. Retrieved From http://search.proquest.com/docview/237239419

Jedel, I., Palmquist, A., Munkvold, R., Goethea, O., Jonasdottira, H., & Erik Olssona, E. (2021). An industry experiment of academic performance and drop-out in gamified distance education. In *5th international GamiFIN conference 2021 (GamiFIN 2021)*, April 7–10, 2021, Finland.

Khalil, M., Wong, J., De-Koning, B., Ebner, M., & Paas, F. (2018). Gamification in MOOCs: A review of the state of the art. In *IEEE global engineering education conference, EDUCON*. https://doi.org/10.1109/EDUCON.2018.8363430

Koivisto, J., & Hamari, J. (2019). The rise of motivational information systems: A review of gamification research. *International Journal of Information Management, 45*, 191–210. https://doi.org/10.1016/j.ijinfomgt.2018.10.013

Kraolis, S. (2021, December 16). *4 awesome examples of gamification in eLearning*. Retrieved April 14, 2022 from https://www.elucidat.com/blog/gamification-in-elearning-examples/ (2021).

KSU. (2022). *Pedagogy for online teaching: Gamification [Website]*. Retrieved April 13, 2022 from https://dli.kennesaw.edu/resources/pedagogyforonlineteaching/gamification.php

Landers, R. N. (2014). Developing a theory of gamified learning. *Simulation & Gaming, 45*, 752–768. https://doi.org/10.1177/1046878114563660

Limantara, N., Hidayanto, A. N., & Prabowo, H. (2019). The elements of Gamification learning in higher education: A systematic literature review. *International Journal of Mechanical Engineering and Technology, 10*(02), 982–991.

McGrath, N., & Bayerlein, L. (2013). Engaging online students through the gamification of learning materials: The present and the future. In H. Carter, M. Gosper and J. Hedberg (Eds.), *Electric Dreams. Proceedings ascilite 2013 Sydney* (pp. 573–577).

Miciuła, I., & Miciuła, K. (2015). Key trends for building business in the Big Data sector, Studia Informatica, 36. *Scientific Papers of the University of Szczecin, 863*, 51–63.

Miciula, I., & Miluniec, A. (2019). Gamification 3.0 for employees involvement in the company. Conference Paper. https://doi.org/10.21125/iceri.2019.2670

Nieto-Escamez, F. A., & Roldán-Tapia, M. D. (2021). Gamification as online teaching strategy during COVID-19: A mini-review. *Frontiers in Psychology, 12*, 648552. https://doi.org/10.3389/fpsyg.2021.648552

Ozcelik, E., Cagiltay, N. E., & Ozcelik, S. N. (2013). The effect of uncertainty on learning in game-like environments. *Computer and Education: An International Journal, 67*, 12–20. Retrieved from http://www.sciencedirect.com/science/article/pii/S0360131513000481

Pappas, C. (2013). *Gamify the Classroom*. Retrieved from http://elearningindustry.com/gamify-the-classroom
Para, A. (2021). The use of gamification in distance education. *e-mentor, 1*(88), 21–29. https://doi.org/10.15219/em88.1499
Renz, J., Hoffmann, D., Staubitz, T., & Meinel, C. (2016). Using A/B testing in MOOC environments. *ACM International Conference Proceeding Series*. https://doi.org/10.1145/2883851.2883876
Richter, G., Raban, D. R., & Rafaeli, S. (2015). Studying gamification: The effect of rewards and incentives on motivation. In T. Reiners & L. Wood (Eds.), *Gamification in education and business* (Vol. 201, pp. 21–46). Springer. https://doi.org/10.1007/978-3-319-10208-5_2
Rohan, R., Pal, D., & Funilkul, S. (2020). Gamifying MOOC's a step in the right direction?: A systematic literature review. *ACM International Conference Proceeding Series*. https://doi.org/10.1145/3406601.3406607
Romero, M. (2017). Competitive, collaborative, and cooperative play. In K. Peppler (Ed.), *The SAGE Encyclopaedia of out-of-school learning* (pp. 115–116). SAGE Publications. https://doi.org/10.4135/9781483385198.n51
Shahri, A., Hosseini, M., Phalp, K., Taylor, J., & Ali, R. (2019). How to engineer Gamification: The consensus, the best practice and the Grey areas. *Journal of Organizational and End User Computing, 31*(1), 39–60. https://doi.org/10.4018/JOEUC.2019010103
Simões, J., Redondo, R. D., & Vilas, A. F. (2013). A social gamification framework for a K-6 learning platform. *Computers in Human Behaviour, 29*(2), 345–353.
Smiderle, R., Rigo, S. J., Marques, L. B., Coelho, J. A. P. M., & Jaques, P. A. (2020). The impact of gamification on students' learning, engagement and behaviour based on their personality traits. *Smart Learning Environments, 7*(3), 1–11. https://doi.org/10.1186/s40561-019-0098-x
Subhash, S., & Cudney, E. A. (2018). Gamified learning in higher education: A systematic review of the literature. *Computers in Human Behaviour, 87*(Feb), 192–206. https://doi.org/10.1016/j.chb.2018.05.028
Sumer, M., & Aydin, C. H. (2018). Gamification in open and distance learning: A systematic review. In *Learning, design and technology* (pp. 1–16). Springer.
Tkaczyk, P. (2010). Grywalizacja. Jak zastosować mechanizmy gier w działaniach marketingowych. Onepress.
Tkaczyk, P. (2012). Gamification. In *How to apply game mechanisms in marketing activities*. Helion.
UGC: University Grants Commission (Open and Distance Learning Programmes and Online Programmes) Regulations, 2020. Retrieved from 5 April, 2022 from https://www.ugc.ac.in/ugc_notices.aspx?id=MjkzMg (2020).
University of Waterloo. (n.d.). *Gamification and game-based learning [website]*. Centre for Teaching Excellence. Retrieved from https://uwaterloo.ca/centre-for-teaching-excellence/
Vaibhav, A., & Gupta, P. (2014). Gamification of MOOCs for increasing user engagement. In *Proceedings of the IEEE international conference on MOOCs, innovation and technology in education*, IEEE MITE 2014. https://doi.org/10.1109/MITE.2014.7020290
Varonis, E. M., & Varonis, M. E. (2015). Deconstructing candy crush: What instructional design can learn from game design. *International Journal of Information and Learning Technology, 32*, 150–164. https://doi.org/10.1108/IJILT-09-2014-0019
Wankel, C., Marovich, M., & Stanaityte, J. (2010). *Cutting-edge social media approaches to business education: Teaching with Linkedin, Facebook, Twitter, Second life, and Blogs*. Information Age Publishing, Inc.
Wood, N. T., Solomon, M. R., Marshall, G. W., & Lincoln, S. (2010). Corporate training goes virtual: A hybrid approach to experiential learning. In W. Ritke-Jones (Ed.), *Virtual environments for corporate education: Employee learning and solutions*. https://doi.org/10.4018/978-1-61520-619-3.ch016
Zichermann, G., & Cunningham, C. (2011). *Gamification by design. Implementing game mechanics in web and mobile applications*. O'Reilly.

Exploring the Potential of Gamification in Omani Higher Education Institutions: A Case Study UTAS Salalah, Oman

Sangeeta Tripathi

1 Introduction

In recent years, the field of education has witnessed significant changes in higher education. Due to frequently emerging technology, the education sector has strengthened and is equipped with several learning tools and platforms (Saini & Al-Mamri, 2019). Artificial intelligence and Gamification have been recognized as powerful tools for curriculum development to improve students' engagement and motivation. As a modern nation, Oman is experiencing one of its fast-growing periods of socio-economic development and is considered at the crossroads of tradition and innovation (Naghdipour, 2021). Understanding the significance of technology in education, the Oman government has also enhanced their elementary education system, education modules, course preparation, reading material, curriculum, and continuous assessment to achieve Oman Vision 2040. However, despite enormous efforts, these advancements are still nascent (Saini & Al-Mamri, 2019). At the level of higher education, the situation is similar.

Moreover, the COVID-19 pandemic has forced many academics to adopt new tools for online engagement. Gamification has been recognized as a significant potential to motivate and engage students in daily classroom learning routines. On the individual level, some academics of Oman higher education institutions have integrated game elements into the learning process to address challenges such as students' disengagement, lack of motivation, and need for critical thinking and problem-solving skills. However, these efforts are still at their beginning stage. This research paper attempts to determine the potential of Gamification and evaluate the awareness and readiness in Oman's higher education institutions to incorporate

S. Tripathi (✉)
University of Technology and Applied Sciences, Salalah, Oman
e-mail: sangeeta@utas.edu.om

A. de Bem Machado et al. (eds.), *Digital Transformation in Higher Education Institutions*, EAI/Springer Innovations in Communication and Computing,
https://doi.org/10.1007/978-3-031-52296-3_6

Gamification, considering factors such as technical capacities, faculty expertise, and student accessibility. The article also tries to analyze awareness and effectiveness of gamified learning in improving students' engagement, motivation, and learning outcomes, comparing it to traditional teaching methods. The following research questions have been devised to reach the research results.

1. How can Gamification be effectively implemented in Omani higher education institutions, specifically at UTAS Salalah, to enhance students' attention and motivation? (Awareness Stage).
2. What gamification strategy can UTAS Salalah employ to generate students' interest and stimulate their desire to engage with educational content? (Interest Stage).
3. How does Gamification influence students' decision-making capability in education activities? (Desire stage).
4. What key factors contribute to successfully implementing Gamification in Omani higher education institutions, focusing on UTAS Salalah, to achieve positive learning outcomes? (Action Stage).

2 Concept of Gamification

Gamification is always mixed with game-based learning. Gamification is derived from game design principles, which include points, different levels of challenges, badges, and leaderboards in a non-game context. It refers to anything other than a game, such as a reward or badges for perfect performance or attendance. It aims to enhance students' motivation or engagement in different contexts, while gamified learning is more directed toward entertaining students (Mora et al., 2015). According to the study by Kim et al. (2009), game-based learning focuses on achieving specific objectives through gameplay. Hence, the game appeals to the students' curiosity. Gamified learning is designed for entertainment, even though it aims at education.

On the other hand, Gamification enhances the learning process and experience without emphasizing the game than learning. It transforms learning into enjoyable activities that encourage students to persist in their studies in the future (Codish & Ravid, 2014). Ceker and Ozdami's (2017) study mentioned Gamification as a non-game-oriented environment that changes into a game environment by using badges and points leaderboards to enhance student engagement and motivation. Gamification enhances the learning process, and learners become persistent in the future. Bozkurt and Genc-Kumtepe (2014) mentioned Gamification as a learning philosophy used in non-game applications. Gamification transforms the traditional classroom experience into an immersive learning journey where students actively participate and take ownership of their learning. By leveraging the innate human inclination toward competition, achievement, and reward, Gamification encourages students to engage more enjoyable and meaningfully with educational content.

3 Need of Gamification in Omani Education System

According to the study by Al-Azawi et al. (2019), there are some difficulties in higher education institutions in maintaining students' engagement and motivation as a group structure. This problem persists in the Gulf region due to language and culture barriers. Oman is also having the same issue. Oman is a modern nation where English is essential to conducting business, science, and technological collaboration, and traveling to English-speaking or non-Arabic countries for work or leisure (Al-Issa, 2006). However, English needs to be taught as a language of communication (Al-Battashy, 1989; Al Toubi, 1998; Al-Issa, 2006). It is worth noting that Oman is one of many countries that face such a problem. According to Ali Al-Issa (2005), Omani students encounter problems with communicative English when they exit high school. The English taught and evaluated in secondary school differs from the English that students need to enter in any English-medium university or college (Saur & Saur, 2001).

According to Al-Issa (2006), Students feel more engaged or motivated when they understand the course content and connect themselves to the classroom learning. When students' interests connect with different classroom activities, it increases students' motivation for better learning. Technology enhances the students' exposure to rich, authentic, and contextual understanding. It enables students to interpret the information and make connections in which the study material is being presented or studied with the natural world functioning. Oman understands the value of education as it transfers people's perception, enables the workforce to contribute to the country's labor market, and supports nation-building. The Ministry of Education emphasized using technology and scientific approaches in teaching and acquiring knowledge to advance society, innovative and productive (Nasser, 2019). Oman government grants over 250 fully or partially funded scholarships to Omani students yearly to pursue higher education in English-speaking countries (Al-Issa (2006). According to Oman National Strategy for Education 2040, Oman's human resources need to be equipped with value and knowledge. It encourages skills to enable a country's workforce to be more productive in the world of knowledge economy. It aims to maintain a pace of development with continuous changes along with maintaining their intrinsic value and national identity, which contribute to the advancement of human civilization (Oman et al. for education, 2040). The education strategy 2040 emphasizes adopting modern pedagogical approaches that include AI and Gamification to enhance students' engagement and prepare graduates for the evolving job market.

Many research studies have highlighted the need and significance of Gamification in Oman's elementary and higher education system. Al-Azawi et al.'s (2019) study found Gamification more suitable for engaging students in such a scenario. Like other countries, Oman recognizes the potential benefits of Gamification in their elementary and higher education. By incorporating gamification elements into their education process, Oman aims to improve student engagement, enhance their learning experience, and foster a more dynamic and interactive learning environment.

Ebrahimi and Alhumairi's (2023) study revealed that game elements increase fifth-grade students' motivation and participation. It will prove a valuable educational tool if carefully selected and implemented. Allmark's (2020) study, "Gamification Implications for Students Motivation," revealed that Gamification and its elements positively impact students' behavior and learning attitude. Instead of affecting direct learning outcomes, Gamification enhances extrinsic and intrinsic motivation toward learning English and other course learning. Al Jahdhami et al.'s (2021) study mentioned that Gamification has become a popular technique for motivating and engaging high school students in different contexts. Here, it is worth noting that Omani students' reluctant attitude and motivation toward English learning results in a lack of care, which impacts students' performance, classroom understanding, quality of the assignment, late submission, and punctuality issues (Al Mahrooqi, 2012). In this scenario, Al Jahdhami et al. (2021) found Gamification a strong tool to address challenges such as disengagement, lack of motivation, and passive learning. Gamification can promote active learning, critical thinking, collaboration, and problem-solving among Omani Students. This recommends gamification implementation widely in the Sultanate.

However, some barriers have also been experienced that discourage the adoption of innovative technology and Gamification in Oman's elementary and higher education. According to the study by Al-Senaidi et al. (2009), five factors create hurdles in adopting technology: lack of equipment, willingness of institutions to support new tools in teaching, lack of IT, lack of confidence, and time constraints.

4 Relevance of the AIDA Model in Implementing Gamification at Oman Higher Education

We live in the digital age where different sorts of technologies are used in education, business, marketing, economic transactions, and public dealing in our daily interactions. For the new generation Z and Alpha (Tafanao et al., 2020), the job market requires new skills beyond arithmetic, reading, and writing. It is mainly characterized by excessive use of technology, which means the workforce needs to be comfortable with various technical tools and applications, critical thinking, problem-solving attitude, public relations, and complex communication skills to solve any issues. In such a scenario, Gamification is touted as an innovative approach to learning and engaging the students (Al-Azawi et al., 2016). Though Gamification was coined in 2002 and again appeared in 2008 (Faiella & Ricciardi, 2015), it frequently started appearing in 2010. The process of implementing Gamification differs from country to country according to their awareness, interest, and desire to adopt. Here, the author found the AIDA model of Elmo Lewis, 1898 very relevant in implementing Gamification in education (Jiangyu Li, 2013) (Fig. 1).

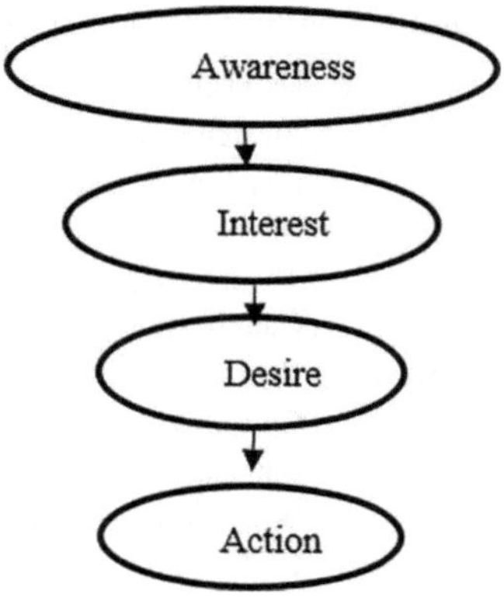

Fig. 1 AIDA Model, self-designed by the author based on AIDA Theory

5 Research Framework

The University of Technology and Applied Sciences (UTAS), one of the leading institutions in Oman, has taken significant strides to incorporate the latest technological approach into its teaching and learning. Salalah UTAS is committed to creating a dynamic and interactive environment to foster active learning, critical thinking, and collaboration between teachers and students. It aims to embark on an exciting journey to revolutionize education by integrating new strategies. To understand the implementation status of Gamification and its future trends, the author uses the AIDA model to design a research framework that describes the status of Gamification at UTAS, Salalah. The AIDA model is adequate for understanding how to engage and motivate students through Gamification. The first stage of the model focuses on raising awareness and getting the attention of the university community. By promoting Gamification through various channels, we can generate interest, desire, and acceptance for its use in teaching and learning. However, only some people aware of Gamification will be interested or motivated to use it. In the second phase, we must focus on the need, utility, and financial support required to motivate the target users. Gamification can address the distinct needs of prospective students and prepare them for the job market. The cognitive phase of awareness can lead to a positive or negative reaction, influencing the disposition toward Gamification. The second and third phases of the AIDA model are empathetic and focus on creating a strong desire to use Gamification. In marketing, this is called the purchase stage, while for technology use, it is called the trial and experiment stage. Using the AIDA model for Gamification in education, we can accomplish definite tasks and engage students more effectively.

Four constructs were devised based on the AIDA model – Awareness, Interest, Desire, and Action and a description for the construct based on the authors' observation (Table 1).

Table 1 Construct and descriptions for the study

Construct	Description
Awareness	Awareness about Gamification in Higher Education
	Elements of Gamification enhance learning attitude among students
	How can Gamification enhance learning in the classroom?
	Awareness status at UTAS Salalah branch to support successful implementation of Gamification in education
Interest	How can gamified learning be helpful for the personalized learning of the students?
	Gamified learning is less trial-and-error learning-based
	Can Gamification in education make assessment easy and limited?
Desire	What has persuaded them to adopt gamified learning in the classroom now and in the future?
	Do they support the assessment of higher education students through Gamification?
	A different element of Gamification that grabs attention and engages students
Action	What are the intricacies and challenges of adopting gamified learning in higher education?
	What strategies can be used to apply gamified learning to motivate and engage our students?
	Willingness to use Gamification in the future

Sources: Created by the author based on the AIDA model and observation

6 Awareness

In the AIDA model, awareness is the first vital element that plays a crucial role in adopting and successfully implementing new technology or applications, which requires proper communication, awareness, the availability of innovative tools and resources. A study by Rogers (2003) emphasized the significant role of awareness before adopting new and innovative technology. His research reveals that awareness about potential benefits equips different stakeholders to make the right decision and help society advance.

Awareness and proper communication create a supportive environment for successfully implementing the technology. According to Bhattacherjee (2001), stakeholders require awareness of emerging technology, resources, training, and supportive information to adopt new technology.

Awareness is also required to overcome challenges and barriers to implementing new applications or technology in different areas. The awareness enables stakeholders to proactively identify and address the challenges (Leonardi, 2011).

7 Interest

In the AIDA model, interest is vital in motivating educators and students to engage with new technology. According to Hidi and Renninger (2006), the interest is closely linked to motivation to adopt technology in learning activities. When instructors are interested in utilizing technology in their classrooms, they explore its potential applications, experience with new tools, benefits, time investment, and efforts in integrating technology into their teaching practices.

Another study by Scherer et al. (2019) reveals that students interested in new technology are likely to actively explore and utilize digital resources, collaborate with peers, and take ownership of their learning. According to Ainley et al. (2002), instructors interested in technology are more likely to design and create engaging learning activities. This can enhance students' participation by motivating them toward learning outcomes.

8 Desire

In the AIDA model, desire is the third significant element. At this stage, users opt for the desire to believe that adopting technology will enhance the learning experience, engage students, and improve learning outcomes. Technology can increase student engagement by facilitating diverse learning preferences. Technology offers interactive, collaborative tools and applications to participate actively in the classroom set-up (Kay, 2012). It also helps in improving learning outcomes. Cheung and Slavin (2013) state that technological integration can lead to higher academic achievements. It can improve critical thinking skills and motivation to retain information for a long time.

9 Action

In the AIDA model, action denotes key strategies and initiatives to embrace new technologies in education. The action can be taken on administrative, academic, or by policymakers. Conducting training and workshops for the instructors' professional development supports and enables them to develop the necessary skills to utilize new technology effectively (Koehler & Mishra, 2009). The action and decision can be taken on the administrative level for comprehensive professional development focusing on pedagogy and technological aspects to empower the university's instructors. Enhancing infrastructure and resources can also be counted under the action category. According to Ertmer et al. (2006), at this stage, academic institutions install and improve adequate infrastructure and robust Internet networking to support a successful implementation of technology-enabled teaching and learning. Leadership vision, guidance, and resources encourage and foster a culture of technology adoption among the instructors and students. Making policies and formulating guidelines to encourage technology integration into the curriculum is crucial for adopting technology (Bebell & Kay, 2010).

10 Methodology

This chapter uses a qualitative research method (DiCicco-Bloom & Crabtree, 2006). The data were drawn from semi-structured interviews (Dadzie et al., 2018) with 15 UTAS academics from Mass Communication, Business, IT, and Math departments. The university instructors are aged between 35 and 45, most of them representing Indian expatriate society from different cities of India, Omani, and Pilipino, living in Salalah, Oman. Snowball sampling has been employed to select respondent instructors from different departments of UTAS. The author found snowball sampling very useful while conducting interviews as the population has specific characteristics that other participants can identify (Cohen & Arieli, 2011). The interview was conducted in three phases. Due to the university's reopening after summer vacation, the first interview phase was conducted on the fourth of September 2023, with four Mass Communication instructors. The second round of interview sessions was done on the fifth of September 2023, with 7 Business, HR, and Marketing instructors. The author encountered a problem while conducting the third phase of the interview. The other four participants showed willingness to participate in the research, but due to the beginning of the semester, they were busy with their students' registration and advising work. Due to time constraints, they requested the author provide them with an online question link so they could fill it out at their convenience. So, the author created a semi-structured interview question link based on previous participants' responses to receive respondents' in-depth insights subjectively. The participant's diverse educational qualifications, including PhD and master's degrees in different disciplines, contribute to a comprehensive understanding of the topic.

11 Discussion and Findings

The present data is drawn from the responses of 15 respondents. The in-depth interview was conducted with each participant, allowing for a detailed exploration of their awareness, and understanding regarding Gamification in higher education. The interview is semi-structured, enabling flexibility while maintaining focus on the research objectives. The answer script is prepared using transcribed verbatim to ensure the accuracy of the data (Rutakumwa et al., 2020). Table 2 shows the sample characteristics of the interviewees.

1. The interview responses revealed that awareness plays a significant role in implementing Gamification in higher education. An average percentage of 1.73 interview responses indicates a basic understanding of Gamification at UTAS, Salalah. Most instructors believe that game-based digital support enhances the attention and motivation of the students.

Table 2 Sample Characteristics = (*n* = *15*)

Characteristic	Freq.	%	Characteristic	Freq.	%
Age			*Nationality*		
35–45	15	100	Omani	4	26.66
46–55	0	00	Indian	10	66.66
56–65	0	00	Pilipino	1	6.67
Gender			*Professional position*		
Female	8	53.33	Lecturer	15	100
Male	7	46.66			
Education					
Master	11	73.33			
PhD	4	26.66			

Dr. Abdulroaf, Mass Communication department, said, "Applying Gamification depends upon the course nature such as English, Math, Interpersonal Skills. These courses have more utility for game learning. However, Gamification has less scope of use in theoretical courses."

2. Exploring the awareness regarding different elements of Gamification, if the UTAS instructors know, 2.73% of interview responses elucidate a shallow knowledge.

Dr. Rathna, an instructor from Business department, described elements as "quizzes, coining the words, and defining the pictures."

Another instructor from the Business department mentioned "badge" while answering other questions, but when the author asked about gamification elements, he said he had not heard about elements.

3. Applying Gamification in higher education is significant, as 2.73 interview responses indicate that Gamification can encourage and motivate average students' learning through active participation and visual elements.

Ms. Saharabano, Business Department, said, "Time is a constraint for theoretical courses; however, gaming applications can enhance engagement and motivation. That is why only one or two classes can be allocated for gamification."

4. Through interviewees' responses, it is evident that awareness regarding Gamification at UTAS, Salalah branch is still early. Some instructors know the basic terms and use Gamification in their classes individually. However, some training and workshop sessions have been conducted during the academic years 2022–2023. That was not sufficient. An average response of 2.73% elucidates that most UTAS instructors believe Gamification benefits specialized courses due to time constraints. Students' responses also matter a lot in implementing Gamification into classroom learning.
5. Exploring the interest regarding the potential gamification strategy employed at UTAS Salalah to generate students' interest and stimulate their desire to engage

with educational content, 2.8% of interviewees' responses indicate that the game is a reason for the student's motivation. Students know that games mean they must participate rather than listen to lengthy lectures and study in class.

According to an instructor from the Business department, "Timespan of the student's attention is limited. Long time listening drifts the students' attention away. The games can retain their attention and motivation, encouraging them to learn."

Characteristics of Gamification include interactive sessions, a participatory environment, and goal-oriented activities that make Gamification interesting for the students.

Ms. Saharabano expressed that Gamification can be used in the classroom as per the requirement of the courses for five to ten minutes. Gamification type and nature should be designed according to the level of the students.

6. Gamification allows the creation of personalized game learning based on the student's unique needs and interests. By using an adaptive algorithm, Gamification can adjust the difficulty and complexity of learning activities for the different needs of the students. This helps them to take the challenges step by step and encourages them to participate with personalized feedback to improve their skills and knowledge. While investigating the perception of UTAS instructors, the author received exciting and diverse opinions regarding gamification personalized learning. Students can take ownership of their progress, and not only the instructors but they can also track their progress record through gamification platforms.

Instructor Butaina, Business department, said, "She prefers to use and create her games based on lessons and students' needs. It helps to engage and motivate students the most. More games, more understanding, enhance the memory."

"Gamification helps to remove stage fear and enhances communication skills," Dr. Rathna, Business Department.

However, some instructors believe that gamification personalized learning is not feasible in higher education. It can be incorporated in the future.

7. Trial and error-based learning supports experimental learning with different strategies, practices, and actions, commonly associated with a problem-solving attitude and hands-on experience. Gamification provides students multiple opportunities to attempt immediate feedback, progress tracking, rewards and achievements, leaderboard, and competition. Its elements encourage students to explore different possibilities, learn from their experience, and gradually improve their skills and knowledge in a particular field. UTAS instructors' response of 1.73% indicates that they agree that Gamification supports trial and error-based learning. However, they also believe that trial-and-error learning cannot be practical and applicable in all cases. It can work only for introductory courses.

Instructor Ricardo Bison, Business Department, said, "Gamification is a mobile-based application which enhances the motivation and participation, but it also hammers the ability to communicate. In the long run, that hinders the ability to learn."

8. Gamification elements can replace traditional grading systems with competency-based assessment models. By defining clear learning objectives, gamified assessments can focus on individual growth and mastery rather than mere grades. While exploring the suitability of Gamification for assessment at UTAS in higher education, 2.26% of UTAS instructors said they do not support Gamification for complete assessment.

Dr. Abdulroaf said, "I am not supporting gamification for assessment. It can be used for classroom activities to motivate and engage the students." Ms. Astloom, Mass Communication, said, "Assessment can affect negatively if the students do not score well in points. It will discourage them." Another UTAS instructor said, "Gamification could not be recommended for a hundred per cent assessment. It is suitable for classroom activities that can be helpful for teachers to access immediate score points to correct students' mistakes."

9. Due to game elements and mechanics involvement, Gamification can change traditional learning into an interactive classroom environment. It can be designed explicitly by defining the clear goal objective of the course into smaller, achievable milestones. Furthermore, points, badges, and leaderboards represent achievement and foster healthy competition among students to strive for excellence. Gamification promotes collaboration and teamwork by incorporating cooperative tasks and challenges that stimulate team spirit and high-order thinking to apply their knowledge to real-life contexts. It also creates continuous growth, storytelling and narration, and technical skills. Immediate feedback helps the students to improve further. An average of 2.4 of interview respondents said that Gamification has the potential to transform the learning experience by harnessing the game elements to make the learning experience more engaging, enjoyable, immersive, and *compelling*.

Ms. Butaina said, "Fun learning makes things easier and motivates students to further learning."

Ms. Wajiha Sayyad, Business department, stated, "Gamification can be used during classroom activities to break the ice instead of assessment because there are different types of learners in the class. Visual learning relates to most students, grabbing their attention significantly and engaging them in continuous learning. Gamification can greatly help the Gulf region, especially in Dhofar, where students face considerable language barriers."

10. The author found many challenges while exploring the initial stage of full-fledged implementation of Gamification at UTAS, Salalah campus. An average of 1.9% of total interview responses indicate that the prime challenge is awareness regarding Gamification and the lack of training/workshops for instructors

and students. In the UTAS Salalah campus, applying Gamification to the courses is an individual choice. That is why it depends on the readiness and willingness of the teacher.

The second challenge is lengthy coursework for specialized courses. Ms. Saharabano, Business department, says, "Time factor, require resources and students' interest interplay a significant role in implementing gamification."

Dr. Mari Anand, Business department, said, "Along with internet availability, students' interest is also significant to adopt any new thing."

Digital Data Report 2023 indicates that Oman had 4.44 million Internet users and a 96.4% Internet penetration rate of the total population in January 2023 (Oman et al., 2023). This indicates that Oman has a sufficient Internet network. However, in terms of Internet quality, it differs.

Ms. Wajiha Sayyad said, "Some classrooms have poor internet connection. Technology is also a problem. Most of the responsibility is upon the teachers. Along with this, availability of Gamification application and, lack of training for teachers and the students are some prominent challenges."

Ms. Astloom said, "Along with awareness, budget is also a challenge for applying gamification at the official level at UTAS, Salalah."

Mr. Ricado Bison expressed many challenges in implementing Gamification at UTAS. He said, "Technology, awareness, and time constraints to learn these applications are the main challenges."

An average of 1.33% of the interview responses indicate that UTAS instructors use gamified learning according to their understanding and nature of the course. They follow different strategies according to their course content, which is not official and consistent across the campus

Dr. Rathna said, "Gamification implementation should start with focusing on school learning. More training and workshops related to gamification should be facilitated on the UTAS campus as part of the teaching and learning process."

Dr. Abdulroaf said, "Gamification cannot be applicable in all courses at the higher education level. The game changes the classroom environment, but implementing gamified learning depends on the course's nature. Visual elements and interactivity help the students to learn fast. It is more applicable to vocabulary and practical courses."

Mr. Ricardo Bison said, "Due to pedagogy, Gamification can be useful to some extent. It would be good to have one official gamification tool or application. Training and workshops must occasionally be conducted to enhance awareness among students and UTAS instructors."

12 Conclusion

The research findings demonstrate the potential benefits of Gamification in higher education. As a case study, the research focuses on UTAS, Salalah branch. The research findings elucidate that awareness is still in the beginning stage. Some

UTAS instructors have recognized the potential benefits of Gamification in higher education and have started applying gamified learning in their classrooms. They found it full of potential to enhance students' engagement and motivation. The UTAS Instructors found it beneficial in overbearing language barriers, improving communication skills, and creating personalized learning. Interactive and participatory classroom sessions have easily transformed traditional lecture-based methods. It is helpful for those students who are shy and not open to classroom discussions. The UTAS instructors found trial-and-error-based learning a significant feature of Gamification. This feature allows students to work on strategies to reach their learning goals. However, the instructors also agree that trial-and-error learning will only be feasible with some of the courses. Regarding assessment, the instructors agreed that Gamification will provide a comprehensive picture of students' progress. However, they raise concerns about gamified assessment's potential adverse effects, such as discouraging students if they do not score well. Addressing these challenges, such as increasing awareness through gamification awareness campaigns, providing adequate training and workshops, and overcoming technological and resource limitations, can pave the path for successfully implementing Gamification at UTAS and other higher education in Oman.

13 Practical Implication

The findings reveal several significant research implications for policy making, practical implementation of Gamification at higher education, and future research direction. The finding indicates a varying awareness of gamification status at the UTAS Salalah branch. Some instructors have a basic understanding and acknowledge the potential benefits of Gamification, while others have limited knowledge. This indicates a need for increasing awareness and training workshops focused on Gamification, its benefits, and practical implementation strategies at the higher education level. The awareness sessions should target both the instructors and students to ensure a comprehensive understanding and use of Gamification in higher education institutions in Oman.

Furthermore, technical support and official gamification guidelines are required to successfully implement Gamification at higher education institutions and UTAS. The governing body of academic institutions or universities can form a committee to establish official guidelines or policies for applying Gamification in different courses. These official guidelines outline the basic principles and best practices of Gamification in higher education. These guidelines can serve as a reference to the instructors and provide a framework for effectively incorporating Gamification into their courses. Necessary application and technological infrastructure are also required to implement Gamification in higher education. The universities or academic institutions can provide robust Internet connectivity, access to appropriate gamification platforms or tools, and technical support for the instructors and students.

The research findings indicate challenges related to financial and budgetary considerations associated with gamification implementation. The Oman government is focusing on the "Oman Vision 2040" and continuously improving Oman's education system through scientific research and technological upgradation. Implementation of Gamification will be very significant in improving the learning experience at the school and college levels. Gamification should start from elementary and higher education together. This action will involve examining the cost-effective gamification application, developing customized applications according to Omani students' needs, and allocating dedicated funds to purchase necessary resources to support gamification efforts.

This chapter draws the attention of policymakers and game developers to design and implement gamification elements specially tailored to theoretical courses at higher education levels. This could involve developing strategies to incorporate game-based digital support or visual elements that align with the nature of theoretical courses, overcoming the perceived time constraints associated with such courses.

As the research findings reveal a need for more awareness at UTAS, the Salalah branch, a collaboration and knowledge-sharing platform by UTAS instructors, can be a significant step in successfully implementing Gamification on the campus. The instructors who have successfully implemented Gamification can share their experience and support their colleagues who are new to Gamification. Initially, gamification implementation can be started on a small scale at UTAS, focusing on selected courses or departments such as English, Math, and other practical courses. This will allow experimentation and help evaluate the effectiveness of Gamification, along with identifying challenges, if any. Further, it can be developed at a large scale.

As the finding reveals that students' feedback is essential, it will be worth collecting and incorporating students' valuable feedback, preferences, experience, and interests with Gamification. This will be very helpful for developing a customized gamification for Omani students. The students' feedback can also be collected through surveys and regular assessments to improve, refine the implementation strategies, and identify areas where additional support is required.

14 Future Research Direction

There is plenty of scope for future research. New research can focus on identifying the specialized courses or subject areas where Gamification is the most suitable and effective. This could involve examining the nature of the courses and the learning outcomes to determine how Gamification can be integrated to enhance students' engagement, motivation, and learning. Further investigation can be done to explore the potential of Gamification in addressing language barriers, particularly in the Dhofar region, where students face considerable challenges in language comprehension.

Another scope of future study is on theoretical courses, challenges of implementing Gamification in higher education, and possible solutions that can help policymakers and design developers to design customized Gamification for different course needs. Future studies can focus on and investigate the effectiveness of Gamification across different disciplines. This could involve examining the impact of Gamification on students' engagement, motivation, and learning outcomes in both practical and theoretical courses.

References

Ainley, M., Hidi, S., & Berndorff, D. (2002). Interest, learning, and the psychological processes that mediate their relationship. *Journal of Educational Psychology, 94*(3), 545.

Al Jahdhami, A. R. M., Gabarre, S., & Gabarre, C. (2021). Gamifying the English language classroom to motivate Omani teenagers. In *Practical perspectives on educational theory and game development* (pp. 54–83). IGI Global.

Al-Azawi, R., Al-Faliti, F., & Al-Blushi, M. (2016). Educational gamification vs. game-based learning: Comparative study. *International Journal of Innovation, Management and Technology, 7*(4), 132–136.

Al-Azawi, R., Joe, S. A., Al-Obaidy, M., & Westlake, J. (2019). The use of gamification technique in agile development methodology. In *Learning Technology for Education Challenges: 8th international workshop, LTEC 2019, Zamora, Spain*, July 15–18, 2019, proceedings 8 (pp. 3–13). Springer International Publishing.

Al-Battashy, S. T. (1989). *Teaching EFL reading and vocabulary in Omani secondary schools*. Unpublished master thesis, University of Southern California, USA.

Al-Issa, A. (2005). An ideological discussion of the place and role of exams in the Omani ELT system. In *The proceedings of redesigning pedagogy: Research, policy, practice*. Center for Research in Pedagogy & Practice, National Institute of Education, Nanyang Technological University.

Al-Issa, A. (2006). Language problems facing Omani learners of English. *ORTESOL, 24*, 19–26.

Allmark, N. (2020). Gamification: Implications for student motivation. In *Oman 19th international ELT conference: Innovative approaches to English language teaching and learning* (pp. 18–29). Sultan Qaboos University.

Al-Mahrooqi, R. (2012). A student perspective on low English proficiency in Oman. *International Education Studies, 5*(6), 263–271.

Al-Senaidi, S., Lin, L., & Poirot, J. (2009). Barriers to adopting technology for teaching and learning in Oman. *Computers & Education, 53*(3), 575–590.

Al-Toubi, S. S. (1998). A perspective on change in the Omani ELT curriculum: Structural to communicative. Unpublished master's dissertation. University of Bristol, England.

Bebell, D., & Kay, R. (2010). One to one computing: A summary of the quantitative results from the Berkshire wireless learning initiative. *Journal of Technology, Learning, and Assessment, 9*(2), n2.

Bhattacherjee, A. (2001). Understanding information systems continuance: An expectation-confirmation model. *MIS Quarterly*, 351–370.

Bozkurt, A., & Genç-Kumtepe, E. (2014). Oyunlaştırma, oyun felsefesi ve eğitim: Gamification. *Akademik Bilişim, 14*, 147–156.

Çeker, E., & Özdaml, F. (2017). What "Gamification" is and what it's not. *European Journal of Contemporary Education, 6*(2), 221–228.

Cheung, A. C., & Slavin, R. E. (2013). The effectiveness of educational technology applications for enhancing mathematics achievement in K-12 classrooms: A meta-analysis. *Educational Research Review, 9*, 88–113.

Codish, D., & Ravid, G. (2014). Personality based gamification-educational gamification for extroverts and introverts. In *In Proceedings of the 9th CHAIS conference for the study of innovation and learning technologies: Learning in the technological era* (Vol. 1, pp. 36–44). The Open University of Israel.

Cohen, N., & Arieli, T. (2011). Field research in conflict environments: Methodological challenges and snowball sampling. *Journal of Peace Research, 48*(4), 423–435.

Dadzie, J., Runeson, G., Ding, G., & Bondinuba, F. K. (2018). Barriers to adoption of sustainable technologies for energy-efficient building upgrade—Semi-structured interviews. *Buildings, 8*(4), 57.

DiCicco-Bloom, B., & Crabtree, B. F. (2006). The qualitative research interviews. *Medical Education, 40*(4), 314–321.

Ebrahimi, R., & Alhumairi, A. (2023, July). Investigating the impact of gamification in education in an International School in Oman: An analysis of students, teachers. In *EdMedia+ innovate learning* (pp. 753–758). Association for the Advancement of Computing in Education (AACE).

Ertmer, P. A., Ottenbreit-Leftwich, A., & York, C. S. (2006). Exemplary technology-using teachers: Perceptions of factors influencing success. *Journal of Computing in Teacher Education, 23*(2), 55–61.

Faiella, F., & Ricciardi, M. (2015). Gamification and learning: A review of issues and research. *Journal of e-Learning and Knowledge Society, 11*(3).

Hidi, S., & Renninger, K. A. (2006). The four-phase model of interest development. *Educational Psychologist, 41*(2), 111–127.

Jiangyu Li, H. Y. (2013, September). An Innovative Marketing Model Based on AIDA: - A Case from E-bank Campus-marketing by China Construction Bank. *iBusiness*. https://doi.org/10.4236/ib.2013.53B010

Kay, R. H. (2012). Exploring the use of video podcasts in education: A comprehensive review of the literature. *Computers in Human Behavior, 28*(3), 820–831.

Kim, B., Park, H., & Baek, Y. (2009). Not just fun, but serious strategies: Using meta-cognitive strategies in game-based learning. *Computers & Education, 52*(4), 800–810.

Koehler, M., & Mishra, P. (2009). What is technological pedagogical content knowledge (TPACK)? *Contemporary Issues in Technology and Teacher Education, 9*(1), 60–70.

Leonardi, P. M. (2011). When flexible routines meet flexible technologies: Affordance, constraint, and the imbrication of human and material agencies. *MIS Quarterly*, 147–167.

Mora, A., Riera, D., Gonzalez, C., & Arnedo-Moreno, J. (2015, September). A literature review of gamification design frameworks. In *2015 7th international conference on games and virtual worlds for serious applications (VS-Games)* (pp. 1–8). IEEE.

Naghdipour, B. (2021). English writing pedagogy at the crossroads: The case of Oman. *Journal of Second Language Writing, 52*, 100815.

Nasser, R. (2019). Educational reform in Oman: System and structural changes. In *Education systems around the world*. Intechopen.

Oman Digital Data Report 2023 can be retrieved from <https://datareportal.com/reports/digital-2023-oman#:~:text=Internet%20use%20in%20Oman%20in,percent)%20between%202022%20and%202023>.

Oman National Strategy for Education, retrieved from < https://www.educouncil.gov.om/downloads/Ts775SPNmXDQ.pdf>.

Rogers, E. M. (2003). *Diffusion of innovations* (5th ed.). Free Press.

Rutakumwa, R., Mugisha, J. O., Bernays, S., Kabunga, E., Tumwekwase, G., Mbonye, M., & Seeley, J. (2020). Conducting in-depth interviews with and without voice recorders: A comparative analysis. *Qualitative Research, 20*(5), 565–581.

Saini, D. K., & Al-Mamri, M. R. S. (2019). Investigation of technological tools used in the education system in Oman. *Social Sciences & Humanities Open, 1*(1), 100003.

Saur, T., & Saur, D. (2001). ESL in Oman's primary, intermediate and secondary schools and how it affects teaching and learning in the intensive English language programme. *Language Centre Forum, 5*(Fall), 9–12.

Scherer, R., Siddiq, F., & Tondeur, J. (2019). The technology acceptance model (TAM): A meta-analytic structural equation modelling approach to explaining teachers' adoption of digital technology in education. *Computers & Education, 128*, 13–35.

Tafonao, T., Saputra, S., & Suryaningwidi, R. (2020). Learning media and technology: Generation Z and Alpha. *Indonesian Journal of Instructional Media and Model, 2*(2), 89–100.

Use of ChatGPT in Higher Education: A Study with Graduate Students

Rogério Costa, Adriana Lage Costa, and Ana Amélia Carvalho

1 Introduction

Almost a year after its release by OpenAI, ChatGPT continues to spark numerous discussions worldwide. There are many challenges and opportunities that artificial intelligence (AI) has been providing to various sectors of society. Still, with a certain degree of incredulity, they seek to understand the potential of these natural language and generative language tools capable of answering questions, performing translations, creating creative text, and facilitating activities in any scientific field in a fast, structured, and effortless way.

However, in the educational setting, criticisms of the new AI model have gained significant prominence and have been growing substantially, without clear guidance from the relevant authorities. In addition to the ethical issues, the growing use of these advanced systems in the classroom, without a careful analysis of their effects and the ability to support and guide their use by teachers and students, has generated concerns.

The use of AI in education is a new, complex, and multifaceted topic that raises a series of questions. On one hand, some see AI as a promising tool to improve teaching and learning (Holmes et al., 2022). They argue that AI can offer

The original version of the chapter has been revised. A correction to this chapter can be found at https://doi.org/10.1007/978-3-031-52296-3_13

R. Costa (✉) · A. L. Costa
Polytechnic Institute of Leiria, Leiria, Faro, Portugal
e-mail: rogerio.costa@ipleiria.pt; adriana.costa@ipleiria.pt

A. A. Carvalho
University of Coimbra, LabTE, CEIs20, Coimbra, Portugal
e-mail: anaameliac@fpce.uc.pt

A. de Bem Machado et al. (eds.), *Digital Transformation in Higher Education Institutions*, EAI/Springer Innovations in Communication and Computing,
https://doi.org/10.1007/978-3-031-52296-3_7

personalization in education by adapting content and teaching strategies to the individual needs of students. Furthermore, AI can provide immediate feedback, assist in performance assessment, and make teaching more efficient. On the other hand, there are significant concerns related to ethics, data privacy, equity (UNESCO, 2021, 2022), excessive dependence on technology, as well as a threat to science and rational thinking (Chomsky, 2023). Some fear that the indiscriminate use of AI could lead to the replacement of teachers and the loss of the human dimension in education. Additionally, there are concerns about how algorithms are trained and whether they can perpetuate or aggravate biases and inequalities.

The divergence of opinions regarding the integration of AI, particularly ChatGPT, into classrooms, fosters a healthy debate of ideas. This variety of perspectives allows for the consideration of both the real benefits and potential risks of the technology. Moreover, this debate provides a timely platform for discussing the importance of implementing appropriate regulations and guidelines for the responsible use of AI in the educational context.

In this context, we emphasize the belief in the most effective approach to mitigate the challenges presented by AI and combat the spread of misinformation: empowering students to use this technology responsibly and thoughtfully. This ongoing dialogue plays a fundamental role in ensuring that AI in education is implemented ethically and beneficially for all parties involved in the educational process.

Researchers are reaching a consensus on the importance of adopting an ethical, comprehensive, and transparent approach when using artificial intelligence in scientific research, integrating it as a fundamental part of the working methodology. This involves declaring the use of AI, documenting the prompts used, and recording the responses obtained (van Dis et al., 2023).

However, when incorporating ChatGPT as a research tool, in addition to the necessary responsibility and critical analysis, researchers must be aware of the tool's limitations regarding the disclosure of data sources, or the textual bodies used in their searches. These factors contribute to a lack of transparency, which is crucial to the ethics of academic research (Lopezosa et al., 2023).

Despite the potential of ChatGPT, researchers should be cautious as there have been many reports of errors and false and biased information, which raises significant concerns about the accuracy and reliability of the information obtained through this tool.

The integration of AI into academic research is undoubtedly of great value but it is not without ethical challenges. Researchers must be diligent in critically evaluating the answers generated by AI, verifying the quality of data sources, and carefully documenting their methodologies. Only in this way can we ensure that AI is an effective ally in academic research while maintaining ethical standards and the integrity of science (van Dis et al., 2023).

UNESCO (2021) recognizes the importance of AI and its impacts on various areas, including education, culture, science, and communication. It has also been promoting discussions and initiatives related to this technology but warns governments of the need for regulations that are crucial to integrate education with a consistent and effective public commitment.

Given the innovative nature of the tool, current research primarily focuses on the use and opinions of teachers and academic scientists. There is a noticeable gap and a scarcity of research in the educational field for a deeper understanding of the impact of ChatGPT in education, with students as fundamental players in the educational process (Strzelecki, 2023).

The situation described above justifies the launch of an exploratory study that aims to identify the usage of ChatGPT by graduate students and their perceptions of this chatbot.

2 ChatGPT: A Generative Natural Language

ChatGPT is the commercial name of a Large Language Model (LLM) product developed by the company OpenAI. Until September 2023, the free 3.5 version of this product was updated with information until January 2022. The latest paid version of ChatGPT 4.0 has access to current web content, as other chatbots such as Bard, launched by Google in 2023, Microsoft Bing Chat, which incorporates ChatGPT 4.0 as its search engine, and LlaMa 2 (Large Language Model Meta AI), launched by Meta.

These LLM are machine learning models centered on natural language processing (NLP). As the name suggests, ChatGPT is based on the GPT-3 architecture, which are the initials of "Generative Pre-trained Transformer." *Generative* means that these models can generate human-like text. They can take a prompt or input text and generate coherent and contextually relevant text as an output. *Pre-trained*, because these models are initially trained on a massive amount of data and then adjusted on specific tasks such as text generation, language translation, code debugging, and question answering. Finally, *transformer* architecture is a type of neural network architecture known for its effectiveness in handling sequential data like text instead of a word of text per se. It is designed to generate human-like text responses in a conversational context, making it a valuable component of AI systems that aim to engage in natural language conversations (Burak, 2023).

When we ask ChatGPT a question, it first processes the input text by tokenizing it (dividing it into separate words) and embedding it (representing each word as a numerical vector). The embedded text is then fed into the transformer, which encodes it and produces a probability distribution for all feasible outputs. This distribution is then used to generate the output response. As we can see, it uses statistical probability to generate models of words and phrases that typically follow any given preceding text (García-Peñalvo, 2023). This process resembles predictive text on a smartphone but has been significantly expanded to produce complete responses instead of isolated words (Haleem et al., 2022; Ouyang et al., 2022).

In summary, the steps involved in ChatGPT's response generation process are:

- Input processing: The user enters a command or question into the ChatGPT text panel.
- Tokenization: The entered text is tokenized into separate words.
- Input embedding: The tokenized text is embedded into a numerical representation.
- Attention encoder-decoder: The transformer encodes the input embedding and produces a probability distribution for all feasible outputs.
- Text generation and output: The transformer generates an output response based on the probability distribution.

ChatGPT can also follow conversations by considering the context of previous messages. This allows it to generate responses that are relevant and coherent within the conversation.

It is noteworthy to acknowledge that generative AI models, such as OpenAI's ChatGPT or Google's Bard, lack real comprehension or consciousness. They generate responses based on language patterns they have learned but have no contextual knowledge of the real world beyond the training data and cannot think or reason like a human. The quality of responses can vary and depends on the quality of training and fine-tuning, as well as the context of the conversation. They choose words in a probabilistic model and they do so at such speed and with such confidence that it gives the user, who is less attentive or has little knowledge, a sense of security about the conversation produced (Lim et al., 2023).

However, as the companies that create these systems highlight, they are not entirely reliable and can generate meaningless or false text, what researchers call "hallucination." The term "hallucination" in AI refers to the model generating false information that sounds plausible (Ji et al., 2023). Probable causes for these "hallucinations" are related to the bias of data, the lack of information from the real world or the limitations of data with which the AI systems were trained. In other words, AI systems can generate unreliable or misleading answers when faced with information for which they have not been directly trained, resulting in a phenomenon of data "hallucination" (Marr, 2023).

Despite experts' efforts in creating models that mitigate this phenomenon, considering the two main factors of "hallucinations": data-related methods and modeling and inference methods (Ji et al., 2023), the production of inaccurate or misleading information leads to eroding user confidence, ethical concerns, impact on decision-making and, ultimately, legal consequences for both users and their creators (Marr, 2023).

Clarity, consistency, and the absence of contradictions in the input prompt, all play an important role in preventing hallucinations. One-way users can actively participate in improving the quality of the results, even though data quality and training are out of their control, is by optimizing the input context.

The fast dissemination and application of AI systems in all areas of knowledge has made the ability to effectively and efficiently interact with AI, one of the most important media literacies. It is therefore crucial that users learn and apply systematic processes to carefully design prompts or inputs, in order to obtain the most accurate results possible and to mitigate the phenomenon of hallucination

(Fui-Hoon Nah et al., 2023). As Gattupalli, Maloy, and Edwards (2023, p. 3) said: "As (…) university professors embark on the journey of integrating AI technology into their classrooms, it is crucial that they understand the significance of this emerging skill of prompt literacy. The ability to create effective prompts will determine the success of AI integration, making it imperative for educators to adapt, personalize, and prioritize this skill."

3 ChatGPT in Education

The adoption of AI tools in education is a growing trend that offers significant opportunities to reflect on their implications for the educational process. However, to enhance the benefits and minimize the challenges associated with this integration, it is essential to deepen the analysis of how AI tools can be effectively adopted in higher education institutions (HEIs). As pointed out by Strzelecki (2023), it is precisely in higher education that ChatGPT offers significant potential to redefine the purpose of assessment, representing a unique opportunity to re-examine not only the underlying objectives of assessment but also the impact that assessment can have on improving learning.

ChatGPT has raised significant concerns in the field of education, particularly with regards to the assessment process, by questioning traditional methods for measuring knowledge, such as tests and exams. AI's ability to generate complex and apparently authentic responses undermines the effectiveness of assessment methods of this nature. This prompts a re-evaluation of how institutions can adapt their assessment strategies to ensure that these adequately reflect student learning in an environment where AI technology is ubiquitous.

Such concerns, mainly the possibility of academic fraud facilitated by access to AI resources such as ChatGPT, have been widely shared by teachers, educational institutions and governments around the world. Recently, UNESCO (2023) released some guidelines for the use of AI in educational institutions, also expressing concern about the excessive speed with which the tool has been implemented in schools and the lack of regulation and control over data protection.

Many teachers and education experts believe students may be tempted to use these tools to generate academic assignments or assessment answers, thereby compromising the integrity of the education system and the validity of assessments. This raises questions about how educational institutions can maintain academic honesty in an environment where AI technology is readily available.

In a recent study analyzing academic literature and news related to AI in higher education, Sullivan et al. (2023) show that academic integrity concerns were discussed more than opportunities to improve teaching and learning with ChatGPT. This finding reinforces the need to focus on the opportunities that surface through AI in education. The movement to curb the use of AI in schools and universities appears to be strongly linked to resistance to change in relation to the adoption of new assessment strategies. To tackle the challenges arising from AI-generated content in student assignments, there is a necessity to revise both assessment techniques and institutional policies (Lo, 2023).

With technology expanding access to knowledge, it is imperative that we develop the ability to deal with uncertainty, evaluate different perspectives and make sense of large amounts of information. This will only be possible if we empower our students to work in collaboration with artificial intelligence, rather than trying to replace it (Gašević et al., 2023).

ChatGPT passed four different examinations at the University of Minnesota Law School (Lo, 2023), proving that it is capable of earning a university degree. This shows that it has introduced new challenges and threats to education. It has to be used with precaution but also with creativity (UNESCO, 2023). Prohibiting its use does not seem to be the best course of action. Research is needed to ensure that its benefits are optimized and its drawbacks are minimized (Lo, 2023).

The responsible and ethical use of ChatGPT is a priority. Teachers reported concerns ranging from plagiarism to incorrect responses, and inaccurate citations (Lo, 2023). ChatGPT plagiarism is not detected in conventional plagiarism detectors, such as Turnitin or iThenticate, but AI-based plagiarism detectors can spot it (Lo, 2023). Lo (2023) identifies several strategies in response to plagiarism issues created by ChatGPT, namely embedding images to exam questions (Susnjack, 2023), using questions that require analysis rather than memorizing, use digital-free assessment formats and checking references during the assessment (Perkins, 2023). Students have to be aware of academic integrity policies and understand the consequences of academic misconduct, as Rudolph et al. (2023) pointed out.

However, it should be noted that, while there are legitimate concerns, ChatGPT also offers opportunities (Sullivan et al., 2023). Educational institutions can use this technology to promote authentic learning by encouraging students to use ChatGPT as a tool to support learning, rather than a solution for academic cheating. This could involve integrating ChatGPT into classroom activities in an ethical way, to promote students' critical thinking and creativity. Educational institutions must harness the full potential of technologies to create discovery-oriented environments, connect students to realities, allow them to explore diverse sources of knowledge, immerse themselves in innovative applications and interact with each other in a more meaningful way (Sullivan et al., 2023).

ChatGPT can also support teachers in several teaching routines, such as teaching preparation and assessment (Lo, 2023). For example, Megahed et al. (2023) asked ChatGPT to generate a course syllabus for an undergraduate statistic course. The authors considered that the teaching suggestions could be adopted with a few adjustments. It can be used also to translate materials into different languages. The chatbot can help teachers in generating exercises, quizzes, and scenarios for students' assessment. However, they have to be checked, as they might not cover all target learning objectives, present basic questions, and have some incorrections.

For students, ChatGTP can answer questions, summarize information, and provide feedback in real time (Lo, 2023). Lo (2023) considers ChatGPT as a virtual tutor to support student's learning. However, it has some inaccuracies that students have to be aware. They should cross-check with other resources, such as books.

Rudolph et al. (2023) mentioned that the chatbot is an aid in improving writing and research skills, but it should not replace critical thinking. Mogali (2023) alerts for well-written texts with inaccurate information. Another issue is that ChatGPT can generate fake bibliographic references (Perkins, 2023; Mogali, 2023; Szabo, 2023).

Given the scarcity of studies reflecting student perspectives, this research aims to examine the use of ChatGPT by higher education students in various contexts and to identify the advantages and disadvantages associated with its use in higher education institutions in Portugal. In addition, the study seeks to identify the activities most frequently carried out by students when using this tool.

4 Methodology

An exploratory study was conducted aimed to characterize the use of ChatGPT by students. We were particularly interested in identifying the perceived advantages and disadvantages by higher education students in Portugal. The study is descriptive and exploratory in nature (Stebbins, 2011; Swedberg, 2020). A single-administration questionnaire was disseminated in some Portuguese higher education institutions.

Following the guidelines of Gil (2008) and Quivy and Campenhoud (2008), the questionnaire was developed in accordance with the research objectives. Certain procedures were adopted to ensure effectiveness and validity in constructing the questionnaire about the use of ChatGPT in an academic context. At each stage of its development, we ensured alignment of the questions with the objectives, categorization by themes, appropriateness of language, and its length. The questionnaire is preceded by an informational letter outlining the details of the study, the estimated time to fill in, ensuring the anonymous and confidentiality of the answers, and the informed consent to collect data. The first part of the questionnaire focuses on the characterization of the respondents, including age, gender, the level of education they are pursuing, the institution of higher education, and whether they are full-time students or part-time student-workers. The second part delves into the context of ChatGPT usages, along with its advantages and disadvantages. The third part presents nine statements with which respondents indicate their agreement or disagreement using a five-point Likert scale. Part 1 and 2 questions are multiple-choice, with one exception in Part 2, which is an open-ended question.

The questionnaire is in line with the Ethical Charter published by the Portuguese Society of Education Sciences (SPCE, 2020). The questionnaire was evaluated by experts, who made suggestions to correct some items and to enhance the clarity of some questions. The questionnaire was tested with a small set of students to verify respondents' comprehension of the items. No issues were detected. The questionnaire was then made available online on the Office Forms platform.

The questionnaire was administered from May to June 2023 in Portugal, and its dissemination was carried out through email by the higher education institutions.

4.1 Participants

A total of 389 responses were obtained from higher education students. Most of the respondents are female (77.1%). They have an average age of 30 years, ranging from 21 to 64 years old. Most of the respondents are master's students (78.4%), and the remaining 21.6% are doctoral students. The majority of the respondents are full-time students (66.8%), while the remaining 33.2% are part-time student-workers. Part-time student-workers come from a wide variety of professions, with teaching being the most common (29%), followed by psychologists (11%). The remaining professions include social workers, researchers, therapists, and pharmacists.

5 Results and Findings

Most students (62.7%) were already familiar with ChatGPT, 32.4% indicated they had heard of it but never used it, and 4.9% did not know what ChatGPT was (Table 1).

Students that were familiar with ChatGPT were asked about the different contexts in which the chatbot was used (Table 2). Most of them (73.4%) have used it in an academic context, followed by usage in a social context (54.9%), and finally in a professional context (28.3%).

From a list of tasks where ChatGPT might have been used (Table 2), the vast majority marked the option of "searching for information" (80.7%), followed by "looking for initial ideas to tackle a topic" (50.4%). In a second group comes "creating activities" (25.8%) and "translation into other languages" (20.1%). In a third group emerges "defining objectives" (16.0%), "creating travel itineraries" (13.9%), and "others tasks" (12.7%) such as correcting and summarizing texts, brainstorming, problem-solving, coding, nutrition, and psychological help. The least selected was answering assessment tests (4.9%).

The most pointed out advantages of using ChatGPT were "easy search" (60.7%), "quick response time" (58.6%), "summarized information" (50.8%), and "help with the initial approach to a topic" (49.2%). Few participants (10.2%) considered the "originality in the text" produced by ChatGPT, and with the same percentage, they did not find "any major advantages" in using it. Only 4.1% of the respondents selected the option "other advantages." They mentioned objectivity of the answers and a different approach about a topic.

Table 1 ChatGPT knowledge and usage

ChatGPT knowledge and usage (N = 389)	*f*	%
I have heard of it and I have used it	244	62.7
I have heard of it and I never used it	126	32.4
I do not know what ChatGPT is	19	4.9

Table 2 Contexts of use, tasks, advantages, and disadvantages of using ChatGPT (N = 244)

Context of use		*f*	%
	Professional	69	28.3
	Academic	179	73.4
	Social	134	54.9
Tasks carried out on ChatGPT			
	Information search	197	80.7
	Answering assessment tests	12	4.9
	Creating activities	63	25.8
	Defining objectives	39	16.0
	Translation into other languages	49	20.1
	Creating travel itineraries	34	13.9
	Looking for initial ideas to tackle a topic	123	50.4
	Other tasks	31	12.7
Advantages of ChatGPT			
	I can't find any major advantages	25	10.2
	Quick response time	143	58.6
	Easy search	148	60.7
	Originality in the text	25	10.2
	Summarized information	124	50.8
	Help with the initial approach to a topic	120	49.2
	Other advantages	10	4.1
Disadvantages of ChatGPT			
	No disadvantages	11	4.5
	No bibliographic sources	150	61.5
	Incorrect information [hallucination]	136	55.7
	General information	93	38.1
	Excessive dependence on ChatGPT	38	15.6
	Biased answers	73	29.9
	Other disadvantages	10	4.9

With regard to the disadvantages perceived by the students, the concerns felt about "incorrect information" (55.7%) and the "lack of bibliographic sources" (61.5%) stand out, which is in line with the academic use that was mostly indicated by the participants. Although in smaller percentages, students also mentioned "general information" (38.1%) and "biased answers" (29.9%), as well as "excessive dependence on ChatGPT" (15.6%). The two extreme options: there are "no disadvantages" (4.5%) and "other disadvantages" (4.9%), show residual value.

Only 9% of the students responded positively to the question whether they "have already used ChatGPT at the request of a teacher." In other words, although students are using artificial intelligence tools in their academic activities, they are still not being encouraged by their teachers to integrate them into their day-to-day academic

Table 3 Category, subcategory, and indicators on the experience of using ChatGPT

Category	Subcategory	Indicators	%
Use of ChatGPT	Experience	Positive	66.8
		Useful	19.3
		Exploratory	9.8
		Negative	4.1

life. Perhaps some teachers are not using it because they have to receive training on artificial intelligence tools (Moura & Carvalho, 2023).

5.1 *Students' Experience of Using ChatGPT*

The analysis of the open-ended question to the questionnaire regarding their experience with the use of ChatGPT followed the procedures used by Bardin (2018) for content analysis. After the initial exploratory analysis of the answers provided by the participants, four indicators were identified regarding their experience: positive, useful, exploratory, and negative (Table 3).

From the indicators extracted from the coding unit carried out and presented in Table 3, most of the participants consider their experience of using ChatGPT to be positive (66.8%). Participants used adjectives such as "spectacular," "incredible," "excellent," "extraordinary," to show their overall positive experience. Useful was indicated by 19.3% of respondents to address topics, organize thoughts, and clarify initial concepts, as well as a beneficial tool to help start writing, find starting points for research, get preliminary insights and quick answers to information, or clarify specific doubts. It was also mentioned by some participants (7.4%) that ChatGPT helped overcoming thinking blocks when elaborating ideas and organizing information, and others (5.3%) referred to ChatGPT's ease of use and its ability to provide direct answers to questions. Although 19.3% consider it useful for a variety of purposes, 11.9% are aware of its limitations, which can be shown by their mentioning of sparse use and constant checking of the information produced by the tool. A conscious and critical approach to using ChatGPT seems to be the key to getting the maximum benefit from the tool, in the opinion of the students.

Some students (9.8%) stated they used ChatGPT in an exploratory way to develop ideas to complement their academic and professional activities. It appears that the exploratory experience on the part of the students focused on obtaining information and answers to various questions, suggestions for summaries, help with research, and understanding concepts.

Of the negative aspects pointed out by 4.1% of the study participants, the limitations in the answers produced by the chatbot stand out. Participants expressed frustration with the ChatGPT answers, mentioning that they were not always as linear or precise as they expected, which ended up being misleading, especially in specific areas. Students also responded that they would not consider the information provided by ChatGPT to be totally reliable and that they would always check the information through other sources. Inconsistency in the information provided, such as references to non-existent or incorrect papers, are obstacles to using the tool.

An analysis of the experiences reported by the students reveals a wide range of uses for this chatbot. It can be seen that the students are curious about how the artificial intelligence behind ChatGPT works. In addition, ChatGPT has become a valuable resource for improving students' writing and organization of ideas. They use it to get word suggestions, grammar corrections, and even to structure texts more effectively. Another interesting aspect is that many students look to ChatGPT for quick answers to their questions and concepts. They see it as a virtual assistant capable of explaining various topics clearly and concisely. This tool has also proved to be a complement to students' academic and professional activities. It helps with research, providing insights into the search for new topics, providing basic information and initial ideas, allowing students to deepen their studies in areas of interest. Finally, students report they use ChatGPT as an information verification tool. They mention the use of the tool to confirm facts, check data, and clarify doubts, which contributes to the accuracy of the information they consume.

The various ways in which this tool is used highlight its versatility. This makes it a possible tool for pedagogical support and cognitive development when well-planned and carefully incorporated into classroom activities. The ability to adapt this technology to students' individual needs and teaching strategies can prove to be a valuable starting point for educators, thus enriching the learning process.

The absence of criteria to ensure a rigorous assessment of the impact of AI on education and to enable teachers to make informed decisions about its implementation in schools has led to uncertainty and questioning about whether this technology really represents a disruptive transformation (Costa, 2023).

Anderson and Rivera-Vargas (2020) point out that education in the twenty-first century has essentially taken on a digital dimension. Therefore, instructing students to effectively, responsibly, and ethically use AI tools, while at the same time enabling them to interact with them in a creative and critical way, is paramount (Costa, 2023; Elkhodr et al., 2023). It is also important to emphasize that ChatGPT should be used as a complementary tool to formal education and cognitive development. Teachers play a key role in guiding students on the appropriate use of this chatbot, promoting critical reflection and discernment when interacting with the information generated by ChatGPT.

5.2 Students' Perception of ChatGPT

To evaluate the students' comprehension and perception of ChatGPT, we conducted an analysis of their agreement and disagreement levels regarding the nine statements presented in Table 4. A total of 370 participants responded to the third section of the questionnaire.

Table 4 Students' perception about ChatGPT

Items	Descriptive statistics and levels of agreement for the scale	Mean	sd	Disagree %	Neutral %	Agree %
1.	ChatGPT's answers are always accurate.	2.38	0.973	**56.5**	30.3	13.2
2.	ChatGPT is connected to the web and searches for information in real-time.	3.43	1.208	20.3	23.2	**56.5**
3.	ChatGPT has been trained in several languages, but its best performance is in English.	3.52	0.911	8.1	**47.8**	44.1
4.	ChatGPT helps me understand information better.	3.31	0.966	15.7	40.5	**43.8**
5.	Artificial intelligence in education enables personalized learning.	3.28	1.031	21.4	37.3	41.3
6.	ChatGPT can replace interaction with teachers and classmates.	1.62	0.885	**84.1**	11.9	4.0
7.	ChatGPT can be used as a tool to enhance learning.	3.56	1	13.8	27.6	**58.6**
8.	I feel comfortable using ChatGPT to get help with all academic tasks.	2.54	1.133	**47.8**	31.6	20.6
9.	ChatGPT helps me retain information better.	2.79	1.037	31.9	45.9	22.2

The value of the calculated Cronbach's alpha coefficient of 0.742 validates the reliability of the scale (Cohen et al., 2018). This means that the questionnaire is a reliable instrument for measuring students' knowledge of ChatGPT.

Given the average values for each of the items, the data indicates that the students have a reasonable knowledge of ChatGPT (Table 4). They believe it is a useful tool for learning, but also have some concerns about its limits. The students are confident that ChatGPT can be used to obtain information, carry out research, and understand complex concepts. They are reluctant that ChatGPT can replace human interaction or be used for all academic tasks.

The highest average score (3.56) was achieved for *ChatGPT as a tool to enhance learning*, with 58.6% of respondents in agreement (item 7). This is closely followed by item 3, *on language proficiency with the best performance in English* (3.52). Item 4, pertaining to helping to understand information better also scores a relatively high fig. (3.31). Item 2, which assesses whether "ChatGPT is connected to the internet and searches for information in real-time," scored an average of 3.43 with 56.5% of participants in agreement. This highlights the participants' limited awareness of the fact that ChatGPT-3.5 operated without Internet connection.[1] Within the subset of items exhibiting the highest average scores, item 5, *personalizing learning with AI*, with an average rating of 3.28, seems to reflect the students' belief in AI's

[1] ChatGPT3.5 had no access to the Web when data was collected.

capacity to tailor learning experiences by adapting content and the teaching pace to students' individual needs.

In contrast, items 1, 8, and 9 display lower average scores, indicative of the uncertainty expressed by students in regards to specific aspects of ChatGPT. Item 1, which pertains to the *accuracy of ChatGPT's responses*, garnered an average rating of 2.38, signifying a degree of doubt. Similarly, item 8, focusing on the confidence in *ChatGPT's assistance with academic tasks*, received an average score of 2.54, and item 9, addressing the *retention of information with ChatGPT's aid*, registered an average of 2.79. It is noteworthy that half of the participants expressed disagreement with ChatGPT's accuracy in providing answers (56.5%) and exhibited skepticism regarding its assistance with academic tasks (47.8%). These statistics underscore a nuanced and balanced evaluation of ChatGPT's overall accuracy by the study's participants.

Lastly, item 6, *ChatGPT can replace interaction with teachers and classmates* yielded the lowest average score (1.62) and the highest percentage of disagreement (84.1%). This starkly indicates that students do not believe in the irreplaceable role of human interaction within the educational context. They perceive ChatGPT cannot provide the same level of support and guidance offered by teachers and peers, underlining the value they attribute to interpersonal engagement in the learning process.

Items 3, 4, and 9 show a high degree of neutrality among students. 47.8% of students answered "neutral" in the context of ChatGPT's proficiency in various languages (item 3), 40.5% in helping to understand information better (item 4), and 45.9% in retaining information (item 9). This neutrality in student responses could be attributed to the relative novelty of such AI systems, along with a certain level of unfamiliarity or uncertainty surrounding them. Furthermore, these findings underscore students' recognition of the limitations associated with ChatGPT's responses.

6 Limitations and Recommendations for Future Research

Although our findings provide important insights into students' use and perspectives about ChatGPT, there are some limitations. First, the participants represent a convenience sample of higher education students from two institutions that accepted the invitation sent by email to participate in this study. As such, these findings cannot be generalized to all students. However, the number of respondents is reasonable.

Second, the questionnaire was short to stimulate respondents, and some topics were not deeply investigated. In future research, we intend to deepen the study of students' use of ChatGPT and their understanding and knowledge about plagiarism.

More studies are needed with students at all levels of education to provide recommendations for using ChatGPT. Studies are also needed with teachers about their use of ChatGPT and whether they are sensitive to the issue of plagiarism, as well as the assessment methods used.

Finally, research also needs to be conducted in institutions about their guidelines and policies for academic integrity and plagiarism prevention (Lo, 2023), as well as about the training provided to staff, students, and teachers. These studies will contribute to making recommendations for using ChatGPT in education.

7 Conclusion

ChatGPT-3.5 is a recent AI tool, free to use with advantages and some disadvantages that have to be acknowledged by its users. It is important to understand how teachers and students are using it to provide useful recommendations. In this study, we focused on higher education students.

The results of this study, conducted in Portugal, show that the majority of students are familiar with ChatGPT and have used it in different contexts, including academic, professional, and social scenarios. The predominant use is related to searching for information and generating initial ideas to address topics. This demonstrates the usefulness of ChatGPT as a tool to support creativity and research. Students also recognize the ease of searching and the speed of response. We can therefore conclude that the ChatGPT language model can be used effectively in the research development process, offering support both in the search for information and in the initial stimulus for approaching study topics. However, students also expressed legitimate concerns, especially regarding the accuracy of the information provided by ChatGPT, the absence of bibliographic sources, and the potential tendency for incorrect or generalized answers. These concerns stem from AI‘s limitations in identifying accurate and relevant information in its vast database. AI often provides answers based on word associations, without fully understanding the context. This can lead to serious errors when students blindly trust the answers generated by AI. Additional concerns include AI‘s limited ability to discern accurate information, its propensity to reproduce biases and even generate false and generalized information. This underscores the importance of using ChatGPT prudently and verifying information through reliable sources.

Notably, only a small number of students reported having used ChatGPT at the request of teachers, suggesting that, despite widespread use, teachers are not yet widely integrating this tool into academic activities. The reasons for this should be explored in future studies.

An analysis of students' perceptions of ChatGPT reveals that they have a reasonable understanding of the chatbot, recognizing both its advantages and limitations. Although they believe that ChatGPT can be a useful tool for improving learning, they also recognize that it cannot completely replace human interaction, especially in the academic context.

In this study, it is clear that the participants recognize the importance of ChatGPT as an auxiliary tool in their academic and professional lives, as in the study conducted by Elkhodr et al. (2023) with undergraduate and postgraduate students. However, we emphasize the need to use Artificial Intelligence (AI) critically and consciously in the educational environment, in accordance with UNESCO (2022, 2023) recommendations. In addition, we highlight the relevance of the teacher's role in the context of working with AI, and particularly ChatGPT.

Based on the results achieved in this study, it is important to consider ChatGPT and other AI tools in academia training, for both students and for teachers. The ethics of its use, as recommended by UNESCO (2022), the prompts literacy (Jacobs & Fisher, 2023; Nyakundi, 2023), and the limitations surrounding ChatGPT are some topics to be approached during training. It is important that teachers be aware of the possibilities of AI tools (Moura & Carvalho, 2023), and that fraud in academia is also dependent on how performance is evaluated (Daumier & Janke, 2019). Forbidding the use of AI tools is not the solution.

Complementary to this study, we identify the need for further research related to students, teachers, and institutions. It is necessary to deepen the understanding of students' use of prompts literacy in the academia and in their private and social life, as well as their ethical behavior, particularly within academic integrity. As only a few students reported having used ChatGPT at the request of teachers, researching teachers'perceptions about the use of AI tools in academia is also needed. Institutions have an important role in providing training for all agents. Training teachers, students, and staff for using AI tools is paramount to ensure adequate preparation for digital education and academic integrity (UNESCO, 2022, 2023).

References

Anderson, T., & Rivera-Vargas, P. (2020). A critical look at educational technology from a distance education perspective. *Digital Education Review, 37*, 208–229. https://doi.org/10.1344/DER.2020.37.208-229

Bardin, L. (2018). Análise de Conteúdo. *Edições, 70*.

Burak, A. (2023). OpenAI ChatGPT, the most powerful language model: An overview. https://relevant.software/blog/openai-chatgpt-the-most-powerful-language-model-an-overview/

Chomsky, N. (2023, Março 8). The False Promise of ChatGPT. *The New York Times*. https://www.nytimes.com/2023/03/08/opinion/noam-chomsky-chatgpt-ai.html

Cohen, L., Manion, L., & Morrison, K. (2018). *Research methods in education* (8th ed.). Routledge.

Costa, A. (2023). O que é ChatGPT e por que a sua utilização tem preocupado as escolas. *De Rerum Natura*. https://dererummundi.blogspot.com/2023/01/o-que-e-chatgpt-e-por-que-sua.html

Daumiller, M., & Janke, S. (2019). The impact of performance goals on cheating depends on how performance is evaluated. *AERA Open, 5*(4).

Elkhodr, M., Gide, E., Wu, R., Darwish, O., Elkhodr, M., Gide, E., Wu, R., & Darwish, O. (2023). ICT students' perceptions towards ChatGPT: An experimental reflective lab analysis. *STEM Education, 3*(2), 70–88. https://doi.org/10.3934/STEME.2023006

Fui-Hoon Nah, F., Zheng, R., Cai, J., Siau, K., & Chen, L. (2023). Generative AI and ChatGPT: Applications, challenges, and AI-human collaboration. *Journal of Information Technology Case and Application Research, 25*(3), 277–304. https://doi.org/10.1080/15228053.2023.2233814

García-Peñalvo, F. J. (2023). La percepción de la Inteligencia Artificial en contextos educativos tras el lanzamiento de ChatGPT: disrupción o pánico. *Education in the Knowledge Society (EKS)*, 24. https://doi.org/10.14201/EKS.31279

Gašević, D., Siemens, G., & Sadiq, S. (2023). Empowering learners for the age of artificial intelligence. *Computers and Education: Artificial Intelligence, 4*, 100130. https://doi.org/10.1016/J.CAEAI.2023.100130
Gattupalli, S., Maloy, R. W., & Edwards, S. A. (2023). Prompt Literacy: A Pivotal Educational Skill in the Age of AI, *College of Education Working Papers and Reports Series. 6*. https://doi.org/10.7275/3498-wx48
Gil, A. C. (2008). Métodos e técnicas de pesquisa social. *Atlas*.
Haleem, A., Javaid, M., & Singh, R. P. (2022). An era of ChatGPT as a significant futuristic support tool: A study on features, abilities, and challenges. *BenchCouncil Transactions on Benchmarks, Standards and Evaluations, 2*(4). https://doi.org/10.1016/J.TBENCH.2023.100089
Holmes, W., Persson, J., Chounta, I., Wasson, B., & Dimitrova, V. (2022). *Artificial intelligence and education a critical view through the lens of human rights, democracy and the Rule of Law*. The Council of Europe.
Jacobs, H. H., & Fisher, M. (2023). Prompt literacy: A key for AI-based learning. *ASCD, 80*(9) https://www.ascd.org/el/articles/prompt-literacy-a-key-for-ai-based-learning
Ji, Z., Lee, N., Frieske, R., Yu, T., Su, D., Xu, Y., Ishii, E., Bang, J., Madotto, A., & Fung, P. (2023). Survey of hallucination in natural language generation. *ACM Computing Surveys, 55*. https://doi.org/10.1145/3571730
Lim, W. M., Gunasekara, A., Pallant, J. L., Pallant, J. I., & Pechenkina, E. (2023). Generative AI and the future of education: Ragnarök or reformation? A paradoxical perspective from management educators. *The International Journal of Management Education, 21*, 1472–8117. https://doi.org/10.1016/j.ijme.2023.100790
Ling, E., Gandolfi, E., & Ferding, R. E. (2023). Helping PreK–12 teachers overcome the fear of artificial intelligence. In R. E. Ferding, R. Hartshorne, E. Baumgartner, R. Kaplan-Rakowski, & C. Mouza (Eds.), *What PreK–12 teachers should know about educational technology in 2023: A research-to-practice anthology* (pp. 71–79). Association for the Advancement of Computing in Education (AACE).
Lo, C. K. (2023). What is the impact of ChatGPT on education? A rapid review of the literature. *Education in Science, 13*(4), 410. https://www.mdpi.com/2227-7102/13/4/410
Lopezosa, C., Codina, L., & Ferran-Ferrer, N. (2023). *ChatGPT Como apoyo a las systematic scoping reviews: Integrando la inteligencia artificial con el framework SALSA*. CRICC. Universitat de Barcelona.
Marr, B. (2023, March 22). *ChatGPT: What are hallucinations and why are they a problem for AI systems | Bernard Marr*. https://bernardmarr.com/chatgpt-what-are-hallucinations-and-why-are-they-a-problem-for-ai-systems/
Megahed, F. M., Chen, Y. J., Ferris, J. A., Knoth, S., & Jones-Farmer, L. A. (2023). How generative AI models such as ChatGPT can be (Mis)used in SPC practice, education, and research? An exploratory study. *arXiv*. arXiv:2302.10916.
Mogali, S. R. (2023). Initial impressions of ChatGPT for anatomy education. *Anatomical Sciences Education*, 1–4.
Moura, A., & Carvalho, A. A. (2023). *Teachers' perceptions of the use of artificial intelligence in the classroom*. Paper Presented at 9th International Conference on Lifelong Education and Leadership for All – ICLEL'2023, 6th–8th September, in Coimbra.
Nyakundi, H. (2023). How to communicate with ChatGPT – A guide to prompt engineering. https://www.freecodecamp.org/news/how-to-communicate-with-ai-tools-prompt-engineering/
Ouyang, L., Wu, J., Jiang, X., Almeida, D., Wainwright, C., Mishkin, P., Zhang, C., Agarwal, S., Slama, K., Ray, A., Schulman, J., Hilton, J., Kelton, F., Miller, L., Simens, M., Askell, A., Welinder, P., Christiano, P. F., Leike, J., & Lowe, R. (2022). Training language models to follow instructions with human feedback. In E. S. Koyejo, S. Mohamed, A. Agarwal, D. Belgrave, K. Cho, & A. Oh (Eds.), *Advances in neural information processing systems* (Vol. 35, pp. 27730–27744). Curran Associates, Inc.
Perkins, M. (2023). Academic integrity considerations of AI large language models in the post-pandemic era: ChatGPT and beyond. *Journal of University Teaching & Learning Practice, 20*, 7.

Quivy, R., & Van Campenhoudt, L. (2008). In Gradiva (Ed.), *Manual de Investigação em Ciências Sociais*.

Rudolph, J., Tan, S., & Tan, S. (2023). ChatGPT: Bullshit Spewer or the end of traditional assessments in higher education? *Journal of Applied Learning & Teaching, 6*, 1–22.

SPCE. (2020). *Carta Ética*. Sociedade Portuguesa de Ciências da Educação. https://www.spce.org.pt/regulacaoeticodeontologia.html

Stebbins, R. (2011). *Exploratory research in the social sciences*. SAGE Publications. https://doi.org/10.4135/9781412984249

Strzelecki, A. (2023). To use or not to use ChatGPT in higher education? A study of students' acceptance and use of technology. *Interactive Learning Environments, 1–14*. https://doi.org/10.1080/10494820.2023.2209881

Sullivan, M., Kelly, A., & McLaughlan, P. (2023). ChatGPT in higher education: Considerations for academic integrity and student learning. *Journal of Applied Learning and Teaching, 6*(1), 31–40. https://doi.org/10.37074/jalt.2023.6.1.17

Susnjak, T. (2023). ChatGPT: The end of online exam integrity? *arXiv*. arXiv:2212.09292v1.

Swedberg, R. (2020). Exploratory research. In E. C. Elman, J. Gerring, & J. Mahoney (Eds.), *The production of knowledge. Enhancing progress in social science*. Cambridge University Press.

Szabo, A. (2023). *ChatGPT a breakthrough in science and education: Can it fail a test?* OSF Prepr.

UNESCO. (2021). *Reimagining our futures together: A new social contract for education*. UNESCO. https://doi.org/10.54675/ASRB4722

UNESCO. (2022). *Recommendation on ethics of artificial intelligence*. UNESCO. https://unesdoc.unesco.org/ark:/48223/pf0000381137

UNESCO. (2023). *ChatGPT and artificial intelligence in higher education*. Quick Start Guide.

van Dis, E. A. M., Bollen, J., Zuidema, W., van Rooij, R., & Bockting, C. L. (2023). ChatGPT: Five priorities for research. *Nature, 614*(7947), 224–226. https://doi.org/10.1038/d41586-023-00288-7

Role of ChatGPT in the Accounting Field

Paolo Biancone and Federico Chmet

1 Introduction

In the era of ever-evolving digital information, technology is profoundly revolutionizing the landscape of accounting and financial management (Secinaro et al., 2021a, b, 2023). Organizations are embracing advanced solutions to enhance the efficiency and accuracy of their financial operations (Koos & Wachsmann, 2023; Kothari, 2023; Kumar et al., 2023). In this context, artificial intelligence (AI) is emerging as a driving force in the digital transformation of accounting, and one of the most notable developments is represented by advanced language models like ChatGPT (Alshurafat, 2023; Street & Wilck, 2023; Wood et al., 2023).

ChatGPT, based on OpenAI's GPT-3.5/4 architecture, represents a significant evolution in conversational AI (Budhwar et al., 2023). This system specializes in interacting with users like human conversations (Dahmen et al., 2023). Its launch on November 3, 2022, has generated considerable interest among academics, researchers, and professionals due to its accuracy and the ability to provide relevant answers to a wide range of questions (Tyson, 2023). Essentially, this artificial intelligence system learns from user questions, and simultaneously, users contribute to the continuous enrichment of its database (Al Ghatrifi et al., 2023; Cao & Zhai, 2023; Dogru et al., 2023).

Its ability to understand natural language and generate contextually relevant responses has sparked a growing interest in various sectors, including accounting. As a result, this paper aims to explore the use of ChatGPT in the accounting domain, highlighting its applications, benefits, and challenges from the perspectives of

P. Biancone · F. Chmet (✉)
Department of Management "Valter Cantino,", University of Turin, Turin, Italy
e-mail: paolo.biancone@unito.it; federico.chmet@unito.it

A. de Bem Machado et al. (eds.), *Digital Transformation in Higher Education Institutions*, EAI/Springer Innovations in Communication and Computing,
https://doi.org/10.1007/978-3-031-52296-3_8

accounting professionals, those in need of financial analysis, and those using ChatGPT to learn accounting (Gordijn & ten Have, 2023).

Traditionally based on manual procedures and specialized software, accounting is undergoing a technological revolution reshaping the daily activities of accounting experts and financial managers. In this chapter, we analyze how and whether ChatGPT can simplify and improve a wide range of accounting processes, from data entry and categorization to preparing financial reports (Agrawal et al., 2022; Heidt, 2023). However, as we explore these applications and their advantages, we must address associated challenges, such as data security and integration with existing accounting systems (Dowling & Lucey, 2023).

A crucial aspect of this study concerns ChatGPT's impact on the accounting profession and the skills required. The role of accounting professionals is evolving from "recording and reporting" to "analysis and consultancy" (Stokel-Walker, 2023), and ChatGPT can serve as an ally in automating routine tasks, enabling experts to focus on more in-depth analysis and strategic consulting (King, 2023).

The rest of the chapter is structured as follows. Section 2 presents the latest advances in the literature about ChatGPT in the accounting field. In sequence, in Sect. 3, we provide the details of the methodology design. Then, Sect. 4 presents the analysis of results, followed by the discussion and implications in Sect. 5. Finally, Sect. 6 points out the limitations and valuable insights for future research studies.

2 Literature Review

A segment of the extant scholarly discourse posits that ChatGPT has the potential to yield unparalleled advantages for individuals, enterprises, organizations, and supply chains, predominantly enhancing productivity and efficiency (Malik et al., 2023) conducted a study that showed that using ChatGPT by low-skilled workers in their tasks can result in a more significant productivity enhancement than skilled personnel. Nevertheless, the ethical utilization of ChatGPT has been subject to several criticisms and concerns, as highlighted by Liebrenz et al. (2023) and Mhlanga (2023). A recent study (Susnjak, 2022) brought attention to ChatGPT as a possible menace to the veracity of online examinations. Furthermore, the research conducted by Gao et al. (2022) examined the ability of ChatGPT to generate medical abstracts. The findings of this study indicated that the tool can generate abstracts that are indistinguishable from those written by humans, posing a challenge for reviewers to discern between information provided by humans and machines. Bouschery et al. (2023) stated that the interplay between Gen-AI and human agents has the potential to enhance human invention. However, the authors also acknowledged certain constraints, hence instigating a discourse surrounding the extent of human involvement within this dynamic.

Recent research in accounting education has significantly emphasized incorporating ChatGPT into accounting courses, hence requiring a departure from conventional instructional approaches. Integrating technology-enhanced learning environments, such as multimedia instructional methods and mobile learning applications, can augment students' understanding and improve their educational achievements (Dogru et al., 2023). Using contemporary information technology tools such as big data and cloud computing has significantly transformed the accounting field, necessitating inventive pedagogical approaches (Wu, 2021). Courses in practical management accounting should prioritize disseminating business-oriented knowledge and provide students with practical experience, thereby aligning them with the dynamic accounting industry. The ideas of blockchain can be efficiently imparted through practical coding activities and storytelling methodologies, contributing to the modernization of accounting education (Guomin, 2019). These insights hold significant value for academics and organizations aiming to improve instruction and equip students with the necessary skills to navigate the constantly evolving accounting industry.

3 Methodology

The methodology employed for this research was the single-interventionist qualitative case study, chosen for its ability to activate in-depth participatory processes and conduct detailed observations (Yin, 2014). Single case studies can be instrumental as they enhance the understanding of how people frame and solve problems in real-world contexts and can serve as design examples for extrapolation-oriented research (Barzelay, 1993).

This approach proved essential in gaining a profound understanding of operational dynamics in the accounting field, considering the impact of digitalization introduction (Miller & Crabtree, 1994).

In the context of the proposed interventionist case study, an active role was adopted in identifying administrative facts, documenting how they were interpreted, and reporting in financial statements using ChatGPT.

The interventionist approach provided an opportunity to thoroughly explore the potentials and challenges associated with using ChatGPT in the accounting domain (Baard & Dumay, 2020; Larrinaga-Gonzalez & Bebbington, 2001). In line with Baard and Dumay (2020), direct involvement of the researcher in the studied environment was required (Lueg & Lu, 2012).

This study focused on using ChatGPT by accounting professionals, investors, and accounting students in the accounting field. The primary activity involved identifying administrative facts and analyzing how they were processed in a financial statement generated through ChatGPT. This process of active engagement allowed for a detailed understanding of the dynamics involved and provided significant contributions for continuous improvement.

4 Results

In the following section, we will present the study's key findings on using ChatGPT in accounting. We will explore ChatGPT's performance in generating financial statements and analyze the implications of the results for accounting professionals, investors, and accounting students. The collected data will provide an overview of ChatGPT's accounting capabilities and prospects.

4.1 ChatGPT – Potential in the Field of Accounting

The primary objective is to understand the potential applications of ChatGPT in the accounting field. We will closely examine its capabilities in generating financial reports and conducting accounting analysis. The question "What might be your applications in the field of accounting?" produced the following response:

"In the accounting field, my applications can be diverse and impactful. Here are several ways in which I can be utilized:

Data Entry and Bookkeeping: I can assist in automating data entry tasks, helping to accurately record financial transactions, categories expenses, and maintain ledgers.

Financial Analysis and Reporting: I can generate financial reports, analyze trends, and provide insights into financial data, aiding in decision-making processes for businesses and individuals.

Budgeting and Forecasting: I can help create budgets by analyzing historical financial data and generating financial forecasts based on different scenarios.

Tax Preparation: I can provide guidance on tax regulations, help identify deductible expenses, and assist in preparing tax documents, making the tax filing process more efficient.

Auditing Support: I can assist auditors by providing access to financial information, aiding in data analysis, and generating reports for audit purposes.

Risk Management: I can help identify potential financial risks by analyzing financial data, highlighting discrepancies, and offering suggestions to mitigate risks.

Financial Advisory: I can provide general financial advice, answer queries about accounting principles, and guide users through financial decision-making processes.

Invoice Generation and Management: I can automate the creation and tracking of invoices, ensuring accuracy and timely payments.

Employee Payroll: I can assist in processing payroll by calculating salaries, taxes, and deductions, and generating pay slips for employees.

Compliance and Regulation: I can help monitor compliance with accounting standards and regulations, reducing the likelihood of errors or discrepancies.

Cost Analysis: I can analyze costs associated with various aspects of a business, aiding in cost optimization efforts.

Financial Research: I can help gather financial data, statistics, and information for research projects or market analysis.

Training and Education: I can be used as a tool for training accounting professionals, offering explanations of concepts, scenarios, and best practices.

Client Communication: I can assist in communicating financial information to clients, helping them understand complex financial concepts in simpler terms.

Automated Responses: I can provide quick responses to common queries, freeing up accountants' time for more complex tasks.

Remember, while I can offer significant support in these areas, critical thinking and human expertise remain essential in accounting. Additionally, regulations and best practices may change, so it's crucial to stay updated with the latest developments in the field.

The impressive range of capabilities exhibited by ChatGPT has led to a decision to delve deeper into the fields of accounting and financial statement preparation. This advanced system, capable of providing accurate and detailed responses to various questions, represents an extraordinary resource for scholars and industry professionals. In accounting, generating financial reports and conducting accounting analyses offers new perspectives on operational efficiency and accuracy.

With a deeper understanding of ChatGPT's potential in the accounting sector, one can anticipate the opening of new horizons for professionals, students, and researchers. This outlines the future of an increasingly automated and data-informed accounting field. To further explore these capabilities, the following question was posed: "How can you assist me in Data Entry and Bookkeeping?" which generated the response:

The answer is: "I can assist you with data entry and bookkeeping tasks in several ways:

Data Input: You can provide me with financial data, such as transaction details, invoices, receipts, and expenses. I can accurately enter these data into spreadsheets, accounting software, or databases, reducing the manual effort required.

Categorization: I can help categorize transactions based on predefined criteria. For example, I can sort expenses into different categories like office supplies, travel expenses, utilities, etc., making it easier to track and analyze spending.

Ledger Maintenance: I can help you maintain digital ledgers by recording credits and debits for each transaction. This ensures that your financial records are organized and up-to-date.

Automation: I can assist in automating repetitive tasks like recurring transactions, making the data entry process more efficient.

Data Validation: I can check the accuracy and consistency of data entries, flagging any potential errors or inconsistencies that might need further review.

Data Formatting: I can format data according to accounting standards and conventions, ensuring that your financial records are presented in a clear and professional manner.

Reporting: I can generate basic financial reports, such as income statements, balance sheets, and cash flow statements, based on the data you've entered.

Integration with Software: If you're using accounting software, I can help integrate data into the software, ensuring that your digital records are always up-to-date.

Data Retrieval: You can ask me to retrieve specific financial information, such as transaction histories, expense summaries, or income reports, making it easier for you to access the information you need.

Data Analysis: I can assist in performing basic data analysis tasks, such as calculating totals, averages, and other key metrics based on the entered data.

Data Security: Your financial data is sensitive, and I can help ensure its security by processing information within the confines of our conversation and not retaining any personal data.

Remember, while I can provide significant support in data entry and bookkeeping, it's important to review the data for accuracy and make any necessary adjustments. Additionally, for complex or critical financial tasks, it's a good practice to consult with a human accounting professionals."

Following the response received, further investigation was conducted to test ChatGPT's ability to represent certain administrative facts within a financial statement. These facts, subjected to analysis, provide a unique opportunity to assess how well the system can translate complex information into clear and understandable accounting terms. This exploration aims to highlight the effectiveness of ChatGPT in the accounting context, providing additional insights into the practical use of this technology for financial statement preparation.

Can you prepare a balance sheet and income statement for the following transactions?

Beta company has a fully paid-up equity of EUR 50,000.

The beta company purchases raw materials for 50,000 euros. The payment is immediate.

The beta company purchases raw materials for 100,000 euros. The shipping costs amount to 1800 euros. The payment is deferred.

On the first of May, Beta purchased a patent for 300,000 euros, and the estimated useful life is 5 years. The payment is deferred.

Beta company sells finished products for a total value of € 600,000. The payment is immediate.

The beta company pays salaries to its employees for a gross amount of 300,000 euros, and tax and welfare deductions amount to 50,000 and 45,000 euros, respectively.

The response is provided below (Figs. 1 and 2).

ChatGPT adds as a response: "Please note that I have provided placeholders for calculations you need to perform based on the given data. Calculate the appropriate cash, accounts receivable, inventory, accounts payable, and depreciation expense values. Additionally, I need to account for potential tax implications and other expenses not mentioned in the transactions. If you are unfamiliar with accounting calculations, seeking assistance from a qualified accountant is recommended." (Tables 1 and 2)

Balance Sheet:

```yaml
Beta Company
Balance Sheet
As of [Date]

Assets:
  Current Assets:
    Cash and Cash Equivalents: [Calculate total cash available]
    Accounts Receivable: [Calculate total accounts receivable]
    Inventory: [Calculate total inventory]
  Non-Current Assets:
    Patent: EUR 300,000

Total Assets: [Calculate total assets]

Liabilities:
  Current Liabilities:
    Accounts Payable: [Calculate total accounts payable]
  Non-Current Liabilities: [No specific liabilities mentioned]

Equity:
  Equity: EUR 50,000

Total Liabilities and Equity: [Calculate total liabilities and equity]
```

Fig. 1 Balance sheet generated by ChatGPT. (Source: Author's elaboration through ChatGPT)

4.2 ChatGPT – Potential for Financial Analysis

The second objective of the research reported here was to identify how ChatGPT can provide valuable support in financial analysis. To this end, a key question was identified and formulated, guiding the entire investigation; it was necessary to put ourselves in investors' perspective, considering that IAS/IFRS standards are designed to meet their ultimate needs. In this role as virtual investors, the following fundamental question was posed to ChatGPT, serving as the starting point for the analysis itself:

Question: "Would you tell me which companies have the highest Dividend Per Share?"

Answer: "I do not have access to real-time financial data or current stock market information as my knowledge cutoff date is September 2021. Dividend Per Share

Income Statement:

```yaml
Beta Company
Income Statement
For the Period Ending [Date]

Revenue:
  Sales Revenue: EUR 600,000

Expenses:
  Cost of Goods Sold:
    Raw Materials: EUR 150,000
  Operating Expenses:
    Employee Salaries: EUR 300,000
    Tax and Welfare Deductions: EUR 95,000
  Depreciation Expense (Patent): [Calculate depreciation expense]

Net Income: [Calculate net income]
```

Fig. 2 Income statement generated by ChatGPT. (Source: Author's elaboration through ChatGPT)

Table 1 Financial statement

Assets	Year X	Equity and liabilities	Year X
Other intangible assets	260.000,00	Equity	50.000,00
Total non-current assets	**260.000,00**	Profit	108.200,00
Cash and cash equivalents	395.000,00	**Total equity**	**158.200,00**
Total current assets	**395.000,00**	Trade payables	401.800,00
		Employee benefit liabilities	45.000,00
		Tax liabilities	50.000,00
Total assets	**655.000,00**	**Total equity and liabilities**	**655.000,00**

Source: Authors' elaboration

Table 2 Income statement

	Year X
Net revenues	600.000,00
Raw materials, supplies and consumables	(151.800,00)
Staff costs	(300.000,00)
Depreciation and amortisation	(40.000,00)
Net profit	108.200,00

(DPS) can change frequently based on a company's financial performance and dividend decisions. To find information on companies with the highest dividend per share at any given time, you can use financial news websites and stock market analysis tools or consult a financial advisor with access to up-to-date financial data. Websites like Bloomberg, Yahoo Finance, Reuters, or dedicated stock market analysis platforms are good places to start your research. They often provide lists and rankings of companies based on various financial metrics, including Dividend Per Share."

While the response provided by ChatGPT is informative, it presents some significant limitations. Firstly, it does not allow the user to conduct further searches or investigations, limiting their ability to explore financial information in detail. Furthermore, even if additional research were possible, access to updated financial data would not be guaranteed. ChatGPT relies on information up to 2021 and cannot provide real-time data.

Another challenge is the need for more information on unlisted companies. ChatGPT primarily focuses on publicly traded companies, while unlisted companies, which constitute a significant part of the business landscape, need to be adequately represented. Collaboration with specialized databases or other external sources would be necessary to access detailed financial data for such companies, making financial analysis more comprehensive and exhaustive.

5 Discussion

In recent times, ChatGPT has received a lot of attention in accounting and financial statement preparation. While this technology clearly advances process automation, it is important to note that its utilization has limitations and substantial hazards. This chapter discusses the findings of a rigorous examination of ChatGPT's attempts to produce a corporate balance sheet, noting the limitations and flaws in the process.

5.1 Categorization Errors

One of the main limitations observed has arisen in the attribution of balance sheet items. Employee benefit and Tax liabilities were mistakenly included in the income statement when they should instead be categorized as liabilities or offset through specific banking operations. This error can lead to distorted financial valuation and an inaccurate representation of the company's finances.

5.2 Absence of Adjusting Entries

Another weakness is the absence of adjusting entries, such as amortization and depreciation. The lack of these transactions can positively impact the balance sheet's accuracy, as it needs to account for the depreciation of company assets over time. These results in the failure to represent actual economic losses and the decrease in the value of assets.

5.3 Absence of Bank Balances and Liabilities

The absence of bank balances and liabilities in the financial position created by ChatGPT represents a significant issue; these are essential data to understand a company's financial situation and ensure that all financial commitments are adequately recorded. Omitting these details can lead to erroneous financial decisions and an incomplete representation of the company's liabilities and obligations.

Our findings highlight that ChatGPT may seem like a promising solution for financial statement preparation. However, it must still be sufficiently developed to guarantee the accuracy and completeness required in a professional or academic context.

From an educational perspective, using ChatGPT in student training could lead to more accurate learning and a more superficial understanding of accounting principles. The errors highlighted here demonstrate that relying exclusively on this technology could provide an inaccurate view of corporate accounting.

From a practitioner's point of view, preparing a financial position is a complex and sensitive process that requires a deep understanding of accounting regulations and company specifics. Unqualified human supervision can lead to severe errors, regulatory misunderstandings, and incorrect financial decisions. This means that the analysis of the results underscores the importance of using ChatGPT only as a supporting tool in balance sheet preparation, with the supervision and guidance of qualified professionals. ChatGPT can simplify some accounting tasks but not replace the human experience and expertise necessary for proper financial management.

5.4 Limitations of ChatGPT in Financial Analysis

ChatGPT is a sophisticated artificial intelligence with enormous potential in financial data processing. However, while delivering credible results for researchers, professionals, and investors, it is critical to recognize its limitations.

5.5 *Obsoleted Data*

One of the primary challenges in using ChatGPT for financial analysis is related to the use of historical data. However, with the current knowledge, the most recent information available was from 2021. This temporal limitation creates a data lag, as financial markets are notoriously volatile and subject to rapid changes. Investors and professionals require up-to-date data to make informed decisions.

5.6 *Lack of Specific Guidance*

Another drawback noted by ChatGPT is the necessity for more precise instruction or directions. Considering the question: "Would you please tell me which firms have the highest Dividend Per Share?" ChatGPT may be required to provide vital information on how to begin research. For example, it does not highlight company classifications such as Dividend Aristocrats, which could be a good beginning point for dividend analysis.

5.7 *Limited Access to Information*

ChatGPT is based on data that is freely available on the Internet. As a result, it is unable to obtain insider information or detailed financial figures from privately held corporations, limiting its capacity to deliver a complete and comprehensive analysis. Many businesses may only publicly publish a portion of their financial information.

5.8 *Exclusive Approach*

It is worth noting that ChatGPT is a tool that can be considered exclusive due to its reliance on Internet access and proficiency in AI usage. As a result, individuals who lack Internet access or are not proficient in AI may be excluded from engaging in financial analysis, raising concerns related to accessibility and equity in the realm of financial decisions and investments.

To conclude, while ChatGPT has the potential to automate certain aspects of financial analysis, it is imperative to acknowledge its limitations. Investors and financial professionals should exercise caution when utilizing this technology and integrate it with other sources of data and analysis to make informed decisions based on the most up-to-date information available. Human oversight and professional expertise remain crucial for accurate and reliable financial analysis.

6 Conclusion

Our chapter explores the utilization of ChatGPT in the accounting industry, for both educational and professional objectives. We have discovered some significant challenges faced by students, professionals, and investors. However, it is important to keep in mind that technology is continually evolving, and the issues that exist currently might be resolved shortly.

In the context of education, it is evident that ChatGPT cannot be reliably used for student training in accounting. The generation of errors and the need for advanced reasoning capabilities in accounting entries limit its usefulness in the educational setting.

For practitioners involved in financial statement preparation, the limitations of ChatGPT are equally evident. In addition to accounting errors, the inability to conduct detailed analyses and comparisons between different scenarios is a significant drawback. Furthermore, the temporal limitation of data up to 2021 and the exclusive focus on publicly traded companies further restrict the utility of ChatGPT in financial analysis and research.

However, it is essential to emphasize that technology is continuously improving. ChatGPT is an example of how AI can be used to automate some accounting tasks, but it is still in development. The challenges and issues highlighted in this article may be addressed in the coming years through improvements in model training and access to more up-to-date and diverse financial data, including data from non-publicly traded companies.

Therefore, while ChatGPT currently exhibits significant limitations in the accounting field, it will become a more reliable and practical resource for students, professionals, and investors. Monitoring and participating in the development of this technology could be crucial to fully harnessing its potential in the financial and accounting domains.

Prospects for future research in this field are diverse and promising. One possible research direction is focused on improving the analytical capabilities of ChatGPT. This could involve the development of more advanced models that allow for more detailed and accurate financial data processing. Additionally, ongoing data updates are crucial, and future studies may explore the integration of ChatGPT with systems that provide real-time financial data to ensure constant information updates.

Another area of research could involve the inclusion of non-publicly traded companies. Currently, ChatGPT primarily focuses on publicly traded companies, but future studies could develop models capable of considering non-publicly traded businesses, perhaps through collaborations with specialized databases.

The interaction between humans and artificial intelligence is crucial, and therefore, future research could explore how to enhance the user interface to make interactions with ChatGPT more intuitive and personalized.

A relevant area of study is financial risk analysis, where ChatGPT could be used to analyze and predict financial risk for companies, contributing to more informed financial decisions.

Integrating machine learning and AutoML techniques could represent another frontier, further improving ChatGPT's performance in financial analysis.

Ethical and regulatory aspects are of growing importance, so future studies could delve into the ethical and regulatory implications of using ChatGPT in accounting, including data privacy management and regulatory compliance.

Training and education represent another potential research field, studying how to effectively use ChatGPT in educating accounting and finance students.

Systematic evaluation of the accuracy and reliability of ChatGPT in financial analysis is crucial, identifying situations where it is most accurate and those where human supervision is necessary.

Finally, the research could examine the industrial adoption of ChatGPT-based solutions in the accounting and finance sector, assessing the benefits and challenges companies may face during implementation. These perspectives represent only some of the many possible directions for future studies on this topic, reflecting the vast potential of Artificial Intelligence in the financial and accounting fields.

References

Agrawal, A., Gans, J., & Goldfarb, A. (2022). *ChatGPT and how AI disrupts industries* (pp. 1–6). Harvard Business Review Press.

Al Ghatrifi, M. O. M., Al Amairi, J. S. S., & Thottoli, M. M. (2023). Surfing the technology wave: An international perspective on enhancing teaching and learning in accounting. *Computers and Education: Artificial Intelligence, 4*, 100144. https://doi.org/10.1016/j.caeai.2023.100144

Alshurafat, H. (2023). *The usefulness and challenges of chatbots for accounting professionals: Application on ChatGPT*. Available at SSRN 4345921. https://papers.ssrn.com/sol3/papers.cfm?abstract_id=4345921

Baard, V. C., & Dumay, J. (2020). *Interventionist research in accounting: A methodological approach*. Routledge.

Barzelay, M. (1993). The single case study as intellectually ambitious inquiry. *Journal of Public Administration Research and Theory, 3*(3), 305–318. https://doi.org/10.1093/oxfordjournals.jpart.a037172

Bouschery, S. G., Blazevic, V., & Piller, F. T. (2023). Augmenting human innovation teams with artificial intelligence: Exploring transformer-based language models. *Journal of Product Innovation Management, 40*(2), 139–153. https://doi.org/10.1111/jpim.12656

Budhwar, P., Chowdhury, S., Wood, G., Aguinis, H., Bamber, G. J., Beltran, J. R., Boselie, P., Lee Cooke, F., Decker, S., DeNisi, A., Dey, P. K., Guest, D., Knoblich, A. J., Malik, A., Paauwe, J., Papagiannidis, S., Patel, C., Pereira, V., Ren, S., …, Varma, A. (2023). Human resource management in the age of generative artificial intelligence: Perspectives and research directions on ChatGPT. *Human Resource Management Journal, 33*(3), 606–659. https://doi.org/10.1111/1748-8583.12524

Cao, Y., & Zhai, J. (2023). Bridging the gap–the impact of ChatGPT on financial research. *Journal of Chinese Economic and Business Studies, 21*(2), 177–191. https://doi.org/10.1080/14765284.2023.2212434

Dahmen, J., Kayaalp, M. E., Ollivier, M., Pareek, A., Hirschmann, M. T., Karlsson, J., & Winkler, P. W. (2023). Artificial intelligence bot ChatGPT in medical research: The potential game changer as a double-edged sword. *Knee Surgery, Sports Traumatology, Arthroscopy, 31*(4), 1187–1189. https://doi.org/10.1007/s00167-023-07355-6

Dogru, T., Line, N., Mody, M., Hanks, L., Abbott, J., Acikgoz, F., Assaf, A., Bakir, S., Berbekova, A., Bilgihan, A., Dalton, A., Erkmen, E., Geronasso, M., Gomez, D., Graves, S., Iskender, A., Ivanov, S., Kizildag, M., Lee, M., et al. (2023). Generative artificial intelligence in the hospitality and tourism industry: Developing a framework for future research. *Journal of Hospitality & Tourism Research*. https://doi.org/10.1177/10963480231188663

Dowling, M., & Lucey, B. (2023). ChatGPT for (Finance) research: The Bananarama Conjecture. *Finance Research Letters*, Elsevier.

Gao, C. A., Howard, F. M., Markov, N. S., Dyer, E. C., Ramesh, S., Luo, Y., & Pearson, A. T. (2022). Comparing scientific abstracts generated by ChatGPT to original abstracts using an artificial intelligence output detector, plagiarism detector, and blinded human reviewers. *BioRxiv*. https://doi.org/10.1101/2022.12.23.521610

Gordijn, B., & ten Have, H. (2023). ChatGPT: Evolution or revolution? *Medicine, Health Care and Philosophy, 26*(1), 1–2. https://doi.org/10.1007/s11019-023-10136-0

Guomin, S. (2019). Reform and exploration of accounting professional practice teaching under the background of artificial intelligence. *IOP Conference Series: Materials Science and Engineering, 563*(5), 052005. https://iopscience.iop.org/article/10.1088/1757-899X/563/5/052005/meta

Heidt, A. (2023). «arms race with automation»: Professors fret about AI-generated coursework. *Nature*. https://europepmc.org/article/med/36693972

King, M. R. (2023). The future of AI in medicine: A perspective from a Chatbot. *Annals of Biomedical Engineering, 51*(2), 291–295. https://doi.org/10.1007/s10439-022-03121-w

Koos, S., & Wachsmann, S. (2023). Navigating the impact of ChatGPT/GPT4 on legal academic examinations: Challenges. *Opportunities and Recommendations*. https://e-journal.unair.ac.id/MI/article/download/45270/25113

Kothari, A. N. (2023). ChatGPT, large language models, and generative AI as future augments of surgical cancer care. *Annals of Surgical Oncology, 30*(6), 3174–3176. https://doi.org/10.1245/s10434-023-13442-2

Kumar, A., Gupta, N., & Bapat, G. (2023). Who is making the decisions? How retail managers can use the power of ChatGPT. *Journal of Business Strategy*. https://www.emerald.com/insight/content/doi/10.1108/JBS-04-2023-0067/full/html

Larrinaga-Gonzalez, C., & Bebbington, J. (2001). Accounting change or institutional appropriation?—A case study of the implementation of environmental accounting. *Critical Perspectives on Accounting, 12*(3), 269–292. https://doi.org/10.1006/cpac.2000.0433

Liebrenz, M., Schleifer, R., Buadze, A., Bhugra, D., & Smith, A. (2023). Generating scholarly content with ChatGPT: Ethical challenges for medical publishing. *The Lancet Digital Health, 5*(3), e105–e106.

Lueg, R., & Lu, S. (2012). Improving efficiency in budgeting–an interventionist approach to spreadsheet accuracy testing. *Problems and Perspectives in Management, 10*(1), 32–41.

Malik, T., Dwivedi, Y., Kshetri, N., Hughes, L., Slade, E. L., Jeyaraj, A., Kar, A. K., Baabdullah, A. M., Koohang, A., & Raghavan, V. (2023). "So what if ChatGPT wrote it?" Multidisciplinary perspectives on opportunities, challenges and implications of generative conversational AI for research, practice and policy. *International Journal of Information Management, 71*, 102642.

Mhlanga, D. (2023). Open AI in education, the responsible and ethical use of ChatGPT towards lifelong learning. In *Education, the responsible and ethical use of ChatGPT towards lifelong learning (February 11, 2023)*. https://papers.ssrn.com/sol3/papers.cfm?abstract_id=4354422

Miller, W. L., & Crabtree, B. F. (1994). Qualitative analysis: How to begin making sense. *Family Practice Research Journal, 14*(3), 289–297.

Secinaro, S., Calandra, D., & Biancone, P. (2021a). Blockchain, trust, and trust accounting: Can blockchain technology substitute trust created by intermediaries in trust accounting? A theoretical examination. *International Journal of Management Practice, 14*(2), 129–145.

Secinaro, S., Dal Mas, F., Brescia, V., & Calandra, D. (2021b). Blockchain in the accounting, auditing and accountability fields: A bibliometric and coding analysis. *Accounting, Auditing & Accountability Journal, 35*(9), 168–203.

Secinaro, S. F., Calandra, D., Roberto, M., & Lanzalonga, F. (2023). Artificial intelligence and healthcare connection: A state-of-the-art discussion among academics and practitioners. In *Incorporating AI Technology in the Service Sector-Innovations in creating knowledge, improving efficiency, and elevating quality of life* (pp. 1–24). Taylor & Francis Group. https://iris.unito.it/handle/2318/1912631

Stokel-Walker, C. (2023). ChatGPT listed as author on research papers: Many scientists disapprove. *Nature, 613*(7945), 620–621.

Street, D., & Wilck, J. (2023). *«Let's have a chat»: Principles for the effective application of ChatGPT and large language models in the practice of forensic accounting*. Available at SSRN 4351817. https://papers.ssrn.com/sol3/papers.cfm?abstract_id=4351817

Susnjak, T. (2022). ChatGPT: The end of online exam integrity? *(arXiv:2212.09292). arXiv*. http://arxiv.org/abs/2212.09292

Tyson, J. (2023). Shortcomings of ChatGPT. *Journal of Chemical Education, 100*(8), 3098–3101. https://doi.org/10.1021/acs.jchemed.3c00361

Wood, D. A., Achhpilia, M. P., Adams, M. T., Aghazadeh, S., Akinyele, K., Akpan, M., Allee, K. D., Allen, A. M., Almer, E. D., & Ames, D. (2023). The ChatGPT artificial intelligence Chatbot: How well does it answer accounting assessment questions? *Issues in Accounting Education*, 1–28.

Wu, Y. (2021). Practical teaching of management accounting course under the background of artificial intelligence and big data. In *2021 international conference on computers, information processing and advanced education (CIPAE)* (pp. 111–114). https://ieeexplore.ieee.org/abstract/document/9610381/?casa_token=ZzX5RPwLkdgAAAAA:Q8Yo_-4iXaLuRDSUmcoOXKCW7GdjpNI72EpsU-KxIM5fEtR5uCNNp4x2Bq4qfYZW0dWAK1u90Ytcog

Yin, R. K. (2014). *Case study research*. Sage.

A Review ChatGPT and the Future of Education

Andreia de Bem Machado, Antonio Pesqueira, and Maria José Sousa

1 Introduction

Since the world of today is undergoing constant change, creativity has become a skill that is required in all vocations and economic sectors. In this context, it is clear that the myriad socio-ecological issues brought on by the digital era have created a variety of difficulties that call for creative solutions across all facets of society. In this regard, artificial intelligence (AI) is an interdisciplinary field that focuses on the study and development of tools and methods used to mimic, amplify, and augment human intelligence (Zhong, 2018a, b). In this artificial intelligence (AI)-related technological scenario, ChatGPT can increase productivity and they suggest that it is likely to offer significant gains in the banking, hospitality, and tourism, and IT sectors, as well as improve business activities like management and marketing (Dwivedi et al., 2023).

As a result, the goal of this project is to use ChatGPT to map the future of education in light of bibliometric analysis. Five sections make up the research. The first section is the introduction, the second is the concepts related to ChatGPT, the third is the technique used in this study, the fifth is the discussion of the study's findings where the research question is addressed, and the sixth is the authors' closing thoughts.

A. de Bem Machado (✉)
Federal University of Santa Catarina, Florianópolis, Brazil
e-mail: andreia.bem@ufsc.br

A. Pesqueira · M. J. Sousa
ISCTE – University Institute of Lisbon, Lisbon, Portugal
e-mail: antonio_pesqueira@iscte-iul.pt; maria.jose.sousa@iscte-iul.pt

A. de Bem Machado et al. (eds.), *Digital Transformation in Higher Education Institutions*, EAI/Springer Innovations in Communication and Computing,
https://doi.org/10.1007/978-3-031-52296-3_9

2 ChatGPT

In its simplest form, ChatGPT is a sophisticated language model that makes use of deep learning strategies to produce responses that are akin to those of a human in response to inputs in natural language. ChatGPT, a part of the family of OpenAI's generative pre-training transformer (GPT) models, has grown in popularity and importance since 2023, securing its place as one of the most widely used language models that is open to the public. ChatGPT can generate contextually appropriate responses and capture the intricacies of human language by using a large text corpus (Dwivedi et al., 2023).

In important ways, artificial intelligence (AI) is used in fields like education, healthcare, and medical research. Because AI can create innovative content and experiences that improve students' learning, it holds the potential to bring about dramatic changes in the educational sector. A change in instructional methods has been sparked by AI technologies like ChatGPT. Artificial intelligence programs like ChatGPT could alter education in the future by giving students individualized learning opportunities that will help them reach their full potential (Peres et al., 2023).

Teachers can then play the part as prompt engineers while students learn how to improve abilities like creativity, communication, and critical thinking. Teachers can create learning experiences that concentrate on higher-order thinking abilities, encourage cooperation, and increase student engagement by utilizing technology. The use of technology by teachers enables them to give pupils immediate feedback and individualized education, allowing students to advance their abilities more quickly and effectively (Short & Short, 2023).

Instead of emphasizing written writings, education is shifting its emphasis to developing students' debate and critical thinking skills. Students are encouraged to express their ideas verbally so that teachers can assess their verbal fluency and communication skills. Numerous possible educational uses for ChatGPT exist, including learning platforms, tutoring systems, individualized instruction, and classroom management. Teachers can use ChatGPT to give their students an interactive learning environment. This may encourage more meaningful interactions between teachers and students and increase student engagement. Additionally, ChatGPT can automate tiresome processes like grading and providing comments (Eager & Brunton, 2023).

More study is required, nevertheless, to determine how students use ChatGPT to produce content and materials for assessments of their learning. Personalized learning experiences, intelligent content generation, and task automation are all advantages for education. These advantages support flexible access to learning journeys and tailored data-based feedback (Lee, 2023; Megahed et al., 2023).

Additionally, ChatGPT might act as a search engine that answers questions immediately, making it simpler for students to conduct research and write papers. It might also be used as a middleman during brainstorming meetings, assisting in topic choice, and launching research initiatives (Jalil et al., 2023).

ChatGPT has been utilized to generate research abstracts from papers from famous publications in a number of investigations. A group of academic scholars and plagiarism detection algorithms examined the final abstracts. It's interesting to note that ChatGPT-generated abstracts scored incredibly high while avoiding the use of conventional plagiarism detection techniques (Dwivedi et al., 2023; Teubner et al., 2023; Tsai et al., 2023).

By examining a massive amount of literature beyond the scope of a student's knowledge, ChatGPT can also reduce the amount of time, effort, and resources used on potentially pointless experiments. The possible creation of virtual professors to support students in their academic endeavors is one key use of ChatGPT in education (Teubner et al., 2023).

For the benefit of both students and professors, ChatGPT might be used to automatically summarize lecture notes and academic discipline histories, easing record-keeping. ChatGPT can be used by instructors and students to dictate and automatically summarize important details. They can also help with learning, memorization, and comprehension by extracting pertinent information from a variety of instructional texts (Tsai et al., 2023).

Additionally, ChatGPT can help with the creation of competence-building laboratories by looking through a large body of case study data to find projects and real-world examples that adhere to learning standards. Additionally, by providing instructions, reminders, setting milestones, and encouraging educational exchanges, it can help students manage their timelines and learning phases.

A recent paper on the advantages of AI for self-management of continuous learning programs claims that ChatGPT can function as a trustworthy conversational agent to elicit information from teachers and then use the learned information for the benefit of the students in the long run (Lee, 2023).

Other important topics that ChatGPT utility cover include data-driven personalization and adaptive learning environments. Student engagement, motivation, and academic success can all be raised by providing personalized educational experiences. Learning that is adapted to each student's needs can narrow knowledge gaps and enhance educational equity (Rospigliosi, 2023).

By providing a source of precise, quick solutions to research queries, ChatGPT can play a larger role as a prospective research tool. This AI model can collect and compile the relevant data from intricate academic databases. Students can now concentrate on using their discoveries by reducing the time-consuming research procedure (Kasneci et al., 2023).

Additionally, by automating tedious administrative procedures, ChatGPT has the potential to improve classroom administration, allowing teachers to concentrate more on pedagogical issues. The model's ability to provide notes for the classroom, reminders, and scheduling milestones could greatly simplify administrative parts of teaching (Lee, 2023).

Additionally, ChatGPT can be used to examine students' educational paths, revealing learning gaps and offering prospective areas for development. Such a feature might facilitate the creation of personalized learning paths and, as a result, more effective student development (Megahed et al., 2023).

The potential advantages of ChatGPT are equally encouraging. ChatGPT could promote medical science by assisting experts in locating potential study subjects. The model may also increase accuracy and efficiency in patient treatment by aiding in clinical and laboratory diagnosis (Nikolic et al., 2023).

Even if these tools show promise, it is crucial to recognize the ethical issues and restrictions connected to ChatGPT and other artificial intelligence models. Credibility problems, plagiarism problems, and authorship rights problems must all be properly addressed and minimized. It's also crucial to take into account the potential biases included in training data as well as the fact that AI models lack complex comprehension and empathy.

A wide range of exciting applications in education are presented by ChatGPT and AI. While it is obvious that these technologies have the potential to revolutionize society, it is necessary to navigate the ethical, practical, and technical issues they bring with them. To fully harness the potential of these technologies for society's advancement, rigorous ethical standards, ongoing research, and judicious use will be required (Megahed et al., 2023).

3 Future of Education Higher Education

3.1 The Future of Higher Education

The world of higher education is evolving at an unprecedented pace, with key influences including technological advancements, demographic shifts, economic necessities, and societal anticipation steering transformation. Presently, an overarching digital revolution, impacting multiple sectors including education, is redefining the essence of higher learning. This wave of change, propelled by digital natives, pressures educational institutions to innovate and evolve (Ghosh et al., 2021).

Digital transformation stands at the core of these shifts, with digital technologies profoundly influencing the higher education landscape. Born or raised in the digital age, today's students, called digital natives, have inherent technological comfort (Jalil et al., 2023). Their expectation of technology-enhanced learning experiences propels educational institutions to innovate and adapt. Universities employ various technology assets, like Learning Management Systems (LMS), online educational platforms, and mobile learning applications, to enrich learning experiences and expand accessibility (Kasneci et al., 2023).

Globalization has intensified competition among Higher Education Institutions (HEIs), resulting in the need for them to evolve. In addition, the shift in pedagogical perspectives toward a constructivist approach implies a move toward a more student-centered approach. It is expected that higher education will place a greater emphasis on skill enhancement than knowledge accumulation (Dwivedi et al., 2023; Kasneci et al., 2023).

As a result, higher education institutions need to embrace innovative approaches to teaching and learning so that they can remain competitive, as well as provide students with the knowledge and skills necessary for success in the modern world (Lyu et al., 2023; Peters et al., 2023). HEIs to succeed these days, it is essential to develop digital literacy skills, problem-solving skills, critical thinking skills, and collaboration skills (Teubner et al., 2023). In addition, HEIs must develop courses that are relevant to the current needs of industry. It is also imperative that they embrace the use of technology to further enhance the learning process (Tsai et al., 2023).

Globalization plays a significant role, with the worldwide dissemination of knowledge and skills propelling universities into an increasingly competitive arena. The ensuing competition compels HEIs to relentlessly innovate to attract students, faculty, and funding (Megahed et al., 2023).

This leads to an increased focus on research output, international collaboration, and globally relevant education. Furthermore, globalization has elevated students' and academics' mobility, fostering a more diverse and inclusive learning environment (Nikolic et al., 2023; Rospigliosi, 2023). The concept of lifelong learning has recently garnered attention, signaling a departure from traditional education confined to early life stages. This idea underpins the paradigm shift in how we view education and its continuous role throughout our lives (Peters et al., 2023; Strzelecki, 2023).

Higher education's global landscape is undergoing substantial transformations, propelled by technological evolution, global economic trends, demographic flux, and shifting societal expectations. These dramatic changes necessitate a thorough examination of existing practices, methodologies, and higher education fundamentals (Dwivedi et al., 2023; Kasneci et al., 2023). On the pedagogical front, a significant evolution is underway in higher education. Learners craft knowledge from their experiences, which is dominant. As such, we observe a discernible shift towards student-centric learning environments that prioritize the development of critical thinking, problem-solving, and collaboration skills, rather than just facilitating knowledge transfer from educator to student (Butt, 2019; Tang, 2000).

This pedagogical evolution is bolstered by digital transformation, as technology-enabled learning environments cater to individual learning preferences and promote proactive engagement (Qadir, 2023). Moreover, the future of higher education underscores the increasing importance of lifelong learning. With rapid technological advancements and the fluid global job market, ongoing learning and skill enhancement is more critical than ever (Jalil et al., 2023; Megahed et al., 2023).

To cater to this demand, higher education institutions offer more adaptable learning routes, online courses, and professional development programs (Rospigliosi, 2023; Tsai et al., 2023). This trend, set to expand in the future, reshapes higher education from being a life stage to a continuous, lifelong endeavor (Qadir, 2023; Teubner et al., 2023). Therefore, higher education's future is influenced by digital transformation, globalization, pedagogical shifts, and lifelong learning. As these trends persist, they will induce further changes in higher education structure, delivery, and perception (Peres et al., 2023).

However, it is essential to consider new technologies' disruptive influence and innovation potential. Recent publications and research works indicate clear growth trajectories in this area (Yang et al., 2023).

Disruption in higher education is not just an inevitability; it's already manifesting itself in various ways. With the advent of Massive Open Online Courses (MOOCs) and online learning platforms like Coursera and EdX, traditional universities no longer monopolize higher education (Short & Short, 2023; Wang et al., 2023).

These online avenues offer formal education alternatives and lifelong learning and professional development, becoming increasingly crucial in today's fast-paced, dynamic world (Kasneci et al., 2023; Nikolic et al., 2023).

In addition, Educational Technological (Edtech) startups offer innovative learning solutions based on cutting-edge technologies such as Artificial Intelligence, Virtual Reality, and Augmented Reality (Lee, 2023; Peres et al., 2023). These advancements widen the learning scope, making education more immersive and personalized. Edtech innovates education by offering personalized learning solutions tailored to the individual learner's needs and interests. AI and virtual/augmented reality can create a more immersive experience that is not possible in a traditional classroom setting. This allows learners to explore and engage with content in ways that are not otherwise possible (Rospigliosi, 2023).

In the evolving landscape of higher education, we'll observe an amplified focus on employability and alignment with labor market needs (Yang et al., 2023). This shift will underscore the critical role higher education plays in preparing students for the ever-evolving professional landscape.

To this end, universities must equip their students with the skills and competencies necessary to succeed in the modern workplace (Lee, 2023; Peres et al., 2023).

Higher education is facing mounting pressures to validate graduate employability and ensure program value proposition (Peres et al., 2023; Short & Short, 2023).

As a result, universities are emphasizing work-integrated learning, providing students with the chance to acquire practical work experiences during their academic journey. Moreover, universities are bolstering career services and initiating industry partnerships to ensure curriculum relevancy and alignment with current job market demands. To further meet the needs of graduates, institutions are increasingly providing career-focused workshops, mentorship programs, and networking events to equip students with the essential skills to succeed in their chosen fields (Qadir, 2023; Teubner et al., 2023).

However, there is a noticeable discrepancy within the literature regarding the intersection of higher education policies, technological advancements like Artificial Intelligence (AI), and funding mechanisms (Alqahtani et al., 2023). Indeed, policy and funding decisions exert a significant influence on higher education trajectory. Institutions are exploring alternative revenue streams, such as commercial activities, philanthropy, and international student fees (Barney et al., 2001).

Market-based policy strategies emphasize competition, performance-based funding, and accountability. These changes are triggering an increased focus on efficiency, quality assurance, and performance metrics within universities. While these alterations have sparked discourse around higher education's commercialization and commodification, they also present opportunities for innovation and societal responsiveness (Eager & Brunton, 2023).

Furthermore, higher education institutions acknowledge their role in addressing global challenges like climate change, poverty, inequality, and social injustice (Hong, 2023). A mounting commitment to sustainable practices is evident not only in campus operations but also in research, teaching, and community engagement (Jalil et al., 2023).

This commitment entails the integration of sustainability into curricula, the promotion of interdisciplinary research on sustainability, the and encouragement of students to become active global citizens (Eager & Brunton, 2023; Lee, 2023).

Higher education, shaped by various forces and trends, will become increasingly digital, global, flexible, and lifelong. This future will also be characterized by a heightened focus on employability, innovation, policy modifications, and social responsibility (Kasneci et al., 2023). As such, higher education institutions that are adaptive, innovative, and responsive are more likely to flourish (Lee, 2023).

Lastly, Artificial Intelligence, particularly ChatGPT, remains a developing area in higher education. Integration of such technologies into education can elevate the learning experience, rendering it more personalized, engaging, and effective (Lyu et al., 2023).

3.2 *Artificial Intelligence Via ChatGPT in Higher Education*

AI technologies, specifically generative models like ChatGPT, are presenting novel opportunities for reshaping the higher education sector. AI-driven education, a rapidly evolving concept, encompasses personalizing learning, providing immediate feedback, and bolstering student engagement (Eager & Brunton, 2023; Lee, 2023).

In this context, AI-enabled chatbots have emerged as pivotal players in the education sector. Chatbots, such as ChatGPT, are increasingly employed for tutoring, administrative assistance, and student interaction. However, current academic literature necessitates a more nuanced exploration of AI's potential and limitations. While AI undoubtedly enhances the learning experience, it also prompts concerns surrounding ethics, privacy, and equity (Hong, 2023; Lyu et al., 2023).

From 2023, AI and adaptive Learning Management Systems (LMS) are expected to play significant roles across various industries, including education. However, this integration raises certain issues such as plagiarism and intellectual property rights that demand comprehensive attention (Lee, 2023; Peres et al., 2023).

AI's influence permeates numerous aspects of higher education, heralding the potential to revolutionize teaching, learning, and administrative tasks. AI is used in developing intelligent tutoring and adaptive learning systems. These systems leverage AI algorithms to assess learner knowledge, identify gaps, and customize learning materials to individual needs, providing a tailored learning experience (Megahed et al., 2023; Qadir, 2023).

Studies suggest that adaptive learning systems can enhance student learning efficacy by addressing weak areas and facilitating self-paced progress. These technologies offer instant feedback, enabling learners to immediately rectify their errors, a feat challenging to scale in conventional classroom settings (Eager & Brunton, 2023; Lee, 2023).

AI also optimizes administrative tasks in higher education. For instance, AI-powered chatbots employed in student services can address routine questions, allowing staff to focus on more intricate tasks (Short & Short, 2023). Furthermore, machine learning algorithms can analyze patterns in expansive datasets to assist decision-making processes such as predicting enrollment numbers, identifying students at risk, and optimizing course schedules (Yang et al., 2023).

One AI technology influencing higher education is OpenAI's GPT-4, also known as ChatGPT. This language prediction model, powered by machine learning, generates human-like text. In higher education, GPT-4 presents potential applications including intelligent tutoring systems offering personalized instruction and feedback and generating educational content. This saves educators time and resources (Rennings, 2000).

The integration of AI in higher education, while bringing myriad benefits, also surfaces a host of ethical and practical concerns, such as data privacy, algorithmic bias, and the digital divide. These issues warrant careful consideration to ensure AI implementation is responsible in the higher education context (Rospigliosi, 2023).

For instance, AI systems, when utilized for determining admissions or grading courses, could be susceptible to an array of biases. If left unaddressed, these biases can engender unfair outcomes, such as preferential treatment of specific student groups or an adverse impact on students from disadvantaged backgrounds. Hence, it becomes vital to identify and rectify any inherent biases to ensure equitable student outcomes (Short & Short, 2023).

In a similar path, data privacy concerns must be managed to safeguard students' personal information against misuse. Effective data protection protocols and stringent privacy policies must be in place to maintain student data integrity and confidentiality (Peters et al., 2023).

Furthermore, the digital divide – the disparity between individuals with and without access to technology – poses a significant challenge. This issue requires adequate attention to prevent AI from intensifying existing inequalities in higher education access. Strategies could include enhancing technology infrastructure in underprivileged areas and developing inclusive policies to ensure equal access to AI-enabled learning resources. Thus, while AI holds immense potential for transforming higher education, its application should be executed in a manner that promotes fairness, transparency, and inclusivity (Eager & Brunton, 2023; Lee, 2023).

As a transformative language model developed by OpenAI, ChatGPT utilizes deep learning techniques to generate contextual, human-like responses. Since 2023, ChatGPT's prominence has surged, primarily due to its capacity to model human language nuances, making it a noteworthy tool in sectors such as education (Hong, 2023; Lyu et al., 2023).

ChatGPT's potential in education is compelling, promising a future where learning is tailor-made, fostering skills development, creativity, and critical thinking. It can also support educators in evolving pedagogical approaches, facilitating a shift towards immediate feedback and personalized instruction. The focus is now on fostering verbal articulation and debate capabilities among students, with less emphasis on traditional written compositions (Strzelecki, 2023). ChatGPT's diverse applications in education range from personalized learning and tutoring to classroom management. By automating administrative tasks such as grading and feedback delivery, it offers an interactive learning environment that encourages meaningful teacher-student dialogues (Nikolic et al., 2023).

An area requiring further exploration is ChatGPT for student content creation and learning evaluations. ChatGPT can enhance educational experiences by personalizing learning, assisting smart content creation, and automating tasks. It can also function as a responsive search engine, simplifying research processes, and facilitating writing assignments for students (Qadir, 2023; Teubner et al., 2023).

Studies have shown that ChatGPT can generate research abstracts that can be reviewed by academic researchers and plagiarism detection software, which evades conventional plagiarism detection. This capability, along with its potential for analyzing extensive literature, can save students significant time and resources.

ChatGPT assisting virtual teachers in their development is exciting. It could generate automated summaries of classroom notes, streamline record-keeping, assist with dictation, extract pertinent information from educational materials, and support comprehension. In competence-building labs, ChatGPT can identify relevant projects by analyzing case study data and facilitate students' learning timeline management (Hong, 2023; Lyu et al., 2023).

ChatGPT presents significant potential for reshaping education. However, further research is warranted to fully understand its implications and leverage its potential effectively.

ChatGPT's capabilities, extending to data-driven personalization and adaptive learning, promise to enhance student engagement and reduce educational inequities. By efficiently dissecting complex academic databases, ChatGPT serves as an efficient research aid, allowing students to focus more on findings application (Peters et al., 2023).

The potential of this AI model for automating routine administrative tasks facilitates more efficient classroom management, enabling educators to prioritize pedagogical concerns. By analyzing students' educational trajectories, ChatGPT can help identify learning gaps, suggesting areas for improvement, thus fostering more efficient student growth (Peres et al., 2023; Short & Short, 2023).

Ethics must not be overlooked, and addressing and mitigating credibility, plagiarism, authorship rights issues, potential biases in training data, and the lack of a nuanced understanding of AI models is crucial. Responsible handling of personal and academic data is of paramount importance as privacy concerns are on the rise. Similarly, auditing AI systems for bias and efforts to mitigate it are crucial (Yang et al., 2023).

The digital divide also poses a significant challenge. As AI becomes integral to higher education, institutions must strive to ensure all students have access to these technologies, to avoid widening the technological divide. ChatGPT's potential applications in higher education are diverse. It could be employed in intelligent tutoring systems to provide personalized instruction. It also can automate responses to routine student inquiries, conserving staff resources (Hong, 2023; Lyu et al., 2023).

3.3 *The Future of Generative AI, ChatGPT 5*

The advent of ChatGPT 4, and the anticipated release of GPT 5, reveal the vast potential of cutting-edge language processing AI in education. These models can generate human-like text, potentially revolutionizing education through personalized learning experiences (Hong, 2023).

GPT 5, the upcoming iteration of the Generative Pre-training Transformer developed by OpenAI, will likely represent a significant leap in natural language processing. Equipped with an advanced understanding of human language and robust problem-solving capabilities, GPT 5 promises more accurate solutions and unprecedented user engagement in the technology industry (Lyu et al., 2023).

Within the educational realm, GPT 5 is drawing significant attention for its capacity to enhance student engagement and serve as a complementary educational tool. Preliminary research indicates that this advanced AI technology could connect with Web 3 and Metaverse realities. This would enable highly personalized learning experiences that stimulate increased levels of student involvement (Tsai et al., 2023).

This potential to generate tailor-made lessons could reshape the educational landscape by fostering a vibrant learning environment that encourages students to delve into and explore their areas of interest. The development of GPT 5 represents an innovative breakthrough in the production of human-like text and advanced natural language processing within intricate environments and formats (Nikolic et al., 2023).

By utilizing an expansive dataset, GPT 5 can generate text that emulates human writing. This positions it as an invaluable asset in natural language processing applications. As such, GPT 5 is poised to significantly contribute to educational advancements. It promises a future of engaging, personalized learning experiences underpinned by cutting-edge AI technology (Wang et al., 2023).

However, GPT 5's potential should be considered alongside data privacy, intellectual property rights, and pedagogical considerations. These are projected to be pivotal areas of future research. The role of educators, quality control, and data privacy are among the primary ethical and pedagogical considerations of ChatGPT in education (Nikolic et al., 2023).

AI technologies like ChatGPT are expected to shape higher education significantly. While promising, these technologies present challenges such as data management, integration of data privacy regulations, and ensuring system interoperability in a Web 3 reality world. Future education must play a crucial role in navigating these challenges and leveraging AI's transformative potential (Wang et al., 2023).

ChatGPT's ability to offer personalized academic support by analyzing individual learning styles and proficiency levels can improve learning outcomes. Acting as a personal tutor, it provides tailored study strategies and constant academic support. Incorporating GPT 5 into education, despite its numerous potential benefits, prompts critical legal and ethical issues relating to data privacy, AI ethics, and potential inherent biases. It is crucial to implement robust data security protocols to safeguard student privacy. This underscores the need for stringent data protection measures within AI-driven education (Nikolic et al., 2023).

AI in education necessitates comprehensive ethical guidelines and regular bias audits to ensure responsible integration. AI developers must deliver algorithms free of bias and adhere to high ethical standards. This reinforces the notion of fairness and responsibility in AI implementation. To ensure adequate protection for students, AI developers must understand data privacy laws. They must also maintain a commitment to ethical algorithms devoid of bias. As such, ensuring compliance with legal and ethical norms is paramount for the successful and responsible integration of GPT 5 into higher education (Wang et al., 2023).

While ChatGPT in education offers exciting opportunities, it underscores the necessity for responsible implementation. Future research should focus on harnessing AI's potential most ethically and effectively (Rennings, 2000).

The challenges and limitations of employing AI in education, especially issues of data privacy and security, warrant careful consideration. As AI models like ChatGPT rely on extensive data for personalization, upholding student privacy rights through stringent data security measures and data anonymization techniques is critical (Eager & Brunton, 2023; Lee, 2023).

While fears of AI replacing educators may exist, it's vital to view AI as a supplemental tool rather than a replacement. AI tools like ChatGPT can never fully replicate educators' empathy and interpersonal relationships. The aim is to augment teaching and learning processes, not replace human educators. Another challenge is AI bias. Training data bias can lead to biased outcomes, perpetuating stereotypes, and misinformation. Therefore, ensuring diverse and balanced training data sets and conducting regular bias audits are essential (Hong, 2023; Lyu et al., 2023). AI incorporation in education requires adequate training for educators, as a lack of digital literacy can pose significant barriers. Educators should be provided with adequate support and professional development opportunities to confidently use these tools (Rospigliosi, 2023).

Assessing the effectiveness of AI tools like ChatGPT in education is critical. Empirical research should be conducted to understand its impact on learning outcomes and student engagement. Feedback from both students and educators should be collected and analyzed for the continuous improvement of these AI tools (Qadir, 2023; Teubner et al., 2023).

ChatGPT holds tremendous potential for revolutionizing higher education, but careful implementation is required. In addition to providing a glimpse into the future of learning, it represents an exciting step in the evolution of education (Wang et al., 2023).

4 Methodology

The integrative review contributes to the systematic visualization of the state of the art (Torraco, 2016) on the subject in the research and its timeline to the level of production by area, avoiding minimization or repetition of studies, or even tending to bias when looking at a specific theme. For this analysis, the research was organized into five phases, namely: problem formulation, definition of research sources, article selection, screening evaluation, and analytical synthesis of results.

The first phase is the research formulation of the problem that guides this study. This will answer the question: What is the future of education with the use of ChatGPT? To answer this question a database search was conducted which started and ended in the month of June 2023.

In the second phase, called definition of the research sources, some criteria for the selection of the research were defined, such as the delimitation of the research base. It was decided to work with the electronic database Web of Science, considered relevant due to the number of abstracts and references indexed in the space with peer review, as well as its impact on the academic area in the interdisciplinary field, which is the study area of this research.

Considering the problem question, the third phase is the selection of articles and conferences. To do this was to delimit the terms or expressions of the research.

The variations of the expressions adopted for the search are presented in a broader context, in the same proposal, because a concept depends on the context to which it is related and its historical trajectory and conceptual analysis. As a basic principle of the research, we chose to insert the terms and expressions in the "Title", "Abstract" and "Keyword" fields. No restrictions of time, language and area of knowledge, or any other restrictions were allowed.

In the fourth phase, the evaluation of the selection, based on the previously defined criteria, totaled 104, indexed papers, which indicated registration from 2022, the first publication, until June 22, 2023, the date that the research was carried out. To analyze the bibliometric data, we used the bibliometrix software because it is the most compatible with the Web of Science database. The bibliometrix package (Bibliometrix, 2022) is an RStudio "library" that provides a set of tools for quantitative research and is recommended for conducting bibliometric analyses (Aria & Cuccurullo, 2017).

5 Results of Bibliometry

As a result of this collection, it was identified that the 104 papers identified in the search were written by 420 authors, linked to 288 institutions from 43 different countries. A total of 271 keywords were used. Table 1 shows the results of this data collection in a general bibliometric analysis.

Table 1 Bibliometric data

Database	Web of science
Search terms	"CHATGPT" AND "EDUCATION"&$$$;.
Search fields	*"title, abstract e key words"*
Total of retrieved works	104
Authors	420
Institutions	288
Countries	43
Author key words (DE)	271
Knowledge area	33
Type of publication	5

The universe of 104 scientific papers makes up the sample for a general bibliometric analysis of publications in the area of ChatGPT and Education, without specific limitations, which allows for a state of the art of the subject, from the database consulted.

5.1 *The Scientific Publications Scenario*

The studies analyzed comprised 104 studies from the Scopus database. In order to appreciate the results in greater depth for the bibliometric analysis, these results were exported to a bibliographic management software called EndNoteWeb. These data provided the organization of relevant information in a bibliometric analysis, such as temporal distribution; main authors, institutions, and countries; type of publication in the area; main keywords and the most referenced papers.

5.2 *Temporal Distribution of the Studies*

Firstly, the temporal distribution of the studies was analyzed, identifying that the publications were very timid in 2022, with only one publication in the year. In 2023, there was a significant growth in the theme with 103 publications on the subject. For better visualization, it was prepared the Fig. 1.

A pioneering work in the area was identified: Open platforms of artificial intelligence in nursing education: tools for academic progress or abuse? by Siobhan O'Connor. The work explains the open platform of artificial intelligence that is OpenAI which is a research institute focused on the development of artificial intelligence in a safe and beneficial way for humanity. The study explains that OpenAI is known for developing cutting-edge AI technologies and conducting research in various areas including natural language processing, computer vision, and robotics. Thus, the study explains that ChatGPT is a large language model trained by

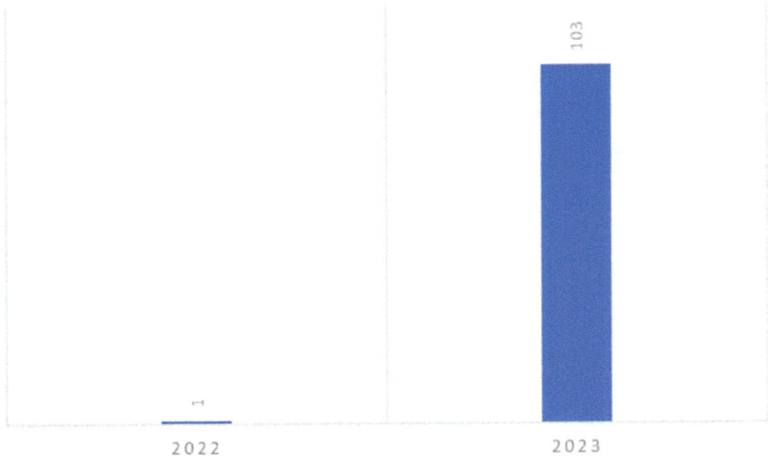

Fig. 1 Times distribution of the papers

OpenAI. It is a machine learning system designed to be able to understand and generate natural human language to assist in a wide range of tasks, such as answering questions and providing information. It is not human but is designed to be able to communicate and interact with people in a person-like manner.

5.3 Distribution of Research Regarding Authors, Institutions, and Countries

Of the 444 citations in the area from different countries, a varied list of authors and institutions that stand out in the research.

When analyzing the country that has the most publications in the area, it can be seen that the United States stands out with an average of 23% of total publications, a total of 115 citations. In second place stands India with 9% of publications, i.e., 45 citations. Figure 2 shows the three countries with the most publications in the research area.

In third place is Australia with 8% of the publications, i.e., 39 citations. In fourth place stands out the UK with 7% citations.

Another analysis carried out is related to the identification of outstanding authors in the area where it was observed that six authors can be called reference in the theme ChatGPT and Education, being considered reference author who has two indexed publications in the area. Seeking to organize these data it was prepared Table 2, with prominent authors in the area, their respective numbers of published articles, institution in which it is affiliated and country.

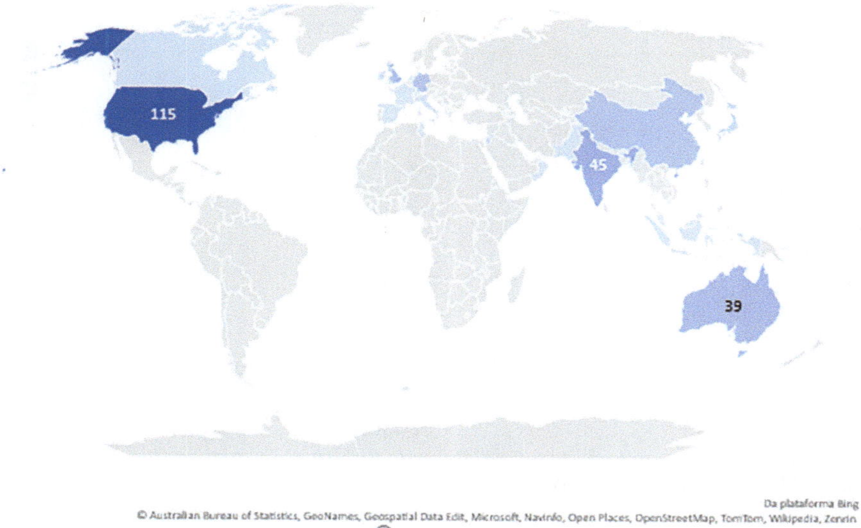

Fig. 2 Country distribution

Table 2 List of authors with the most publications with their affiliations and country

Authors	Number of publications	Affiliation	Country
O'Connor, Siobhan	2	University of Manchester	United Kingdom
Sauerland, Uli	2	Leibniz Zentrum Allgemeine Sprachwissenschaft (ZAS)	Germany
Agerstrand, Marlene	2	Stockholm university	Sweden
Rillig, Matthias C.	2	Freie Universität Berlin	Germany
Sallam, Malik	2	The University of Jordan	Jordan
Arne Pettersen Gould, Kenneth	2	Universitetet I Stavanger	Norway

Based on Table 2 and Fig. 2, it can be seen that the countries with the highest number of publications are not necessarily where the authors with the highest number of publications are found. For example, the United States, which stands out as the country with the most publications in the area, does not appear when analyzing the countries, considering the prominent authors. Germany, on the other hand, which appears in fifth place in the general rank, has two authors, Sauerland, Uli, and Rillig, Matthias C. with two publications. This allows us to infer that where there is a higher concentration of publications, these cannot be considered reference authors in the area. The publication is concentrated in a large number from a diversity of authors.

The most cited authors globally were: ChatGPT Cgpt And O'connor S, with 15 citations in the area, as illustrated in Fig. 3.

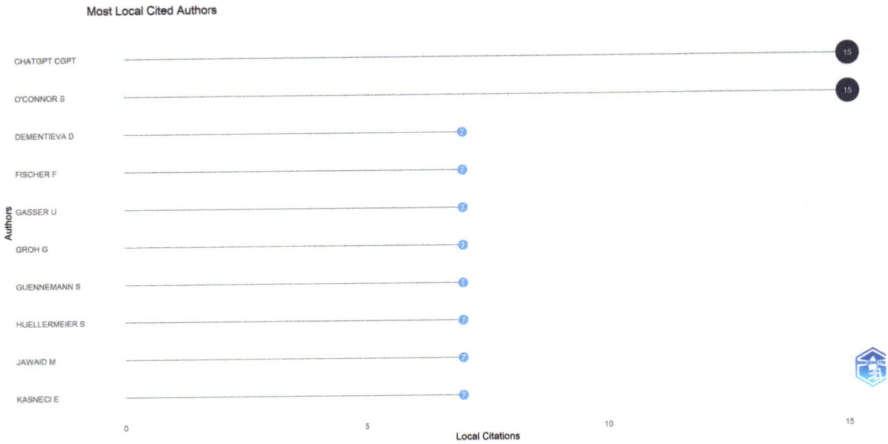

Fig. 3 Most cited authors globally

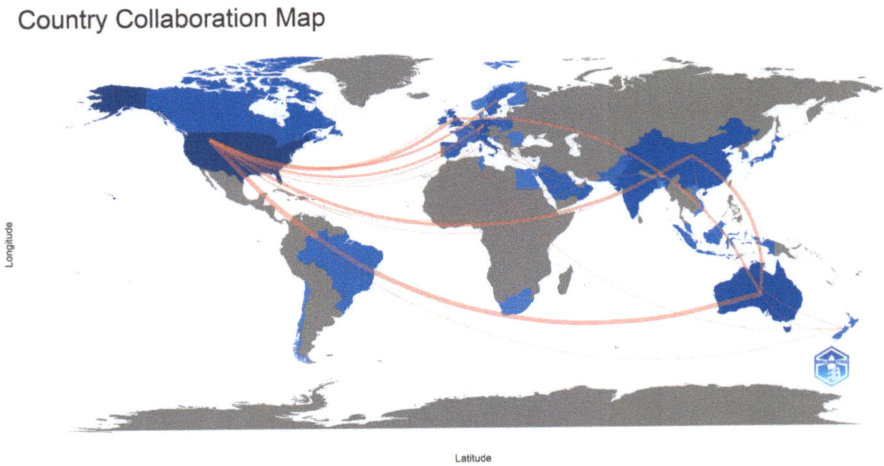

Fig. 4 Spatial distribution and relationships of publications on ChatGPT and Education

Figure 4 presents the intensity of publication by country and the relationship established between them, through citations between published works, the continents: America, Africa, Asia, Europe, Oceania, and Antarctica as shown in the following image.

The documents analyzed were published in 72 different journals, and among the total six studies, that is, (9%) were published in only one journal "Cureus Journal Of Medical Science", according to Fig. 5, which presents the 10 scientific sources with the most publications on the topic "ChatGPT and Education."

From the general survey, it was also possible to analyze the type of document research in the area of ChatGPT and Education. It is noticed that the publications are concentrated in articles in periodicals with 48% of the total number of early

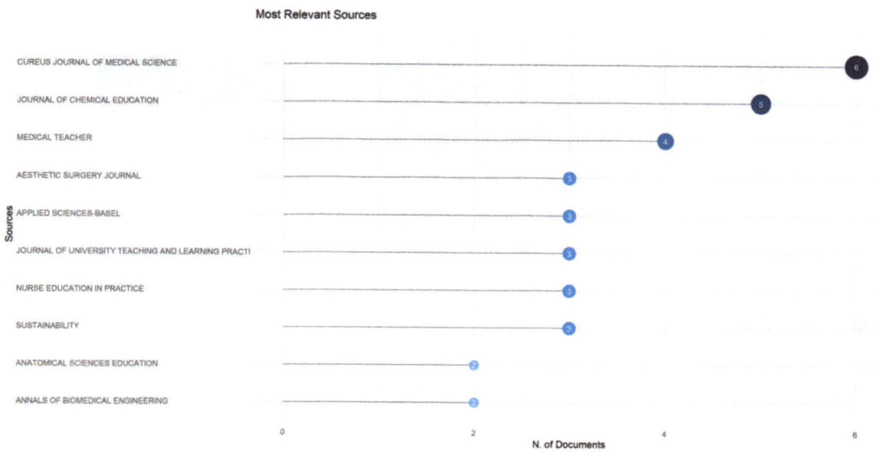

Fig. 5 Scientific sources with most publications

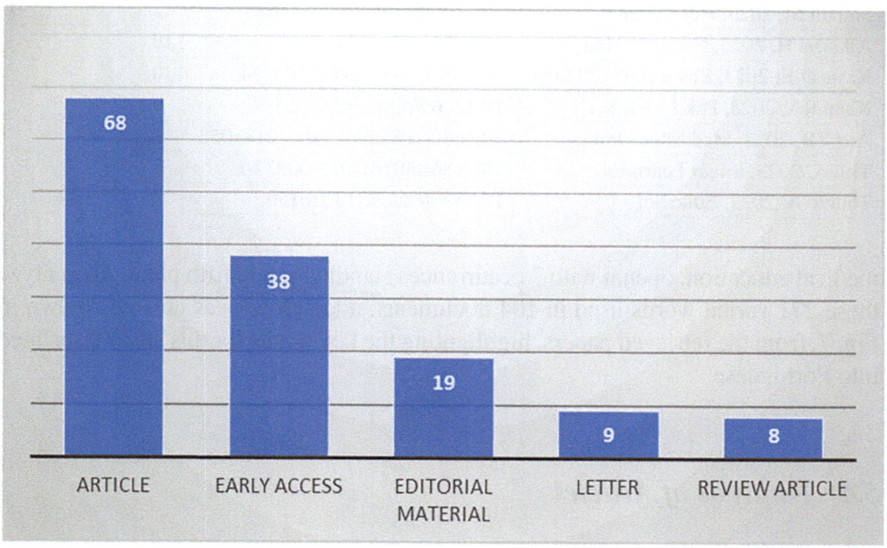

Fig. 6 Distribution of publications by type of journal

access with 27% of the publications. In total, there are five categories of indexations marked, as shown in Fig. 6.

From the bibliometric analysis, based on the workgroup retrieved from the Web Of Sience (WOS) database, it was possible to identify a total of 271 different keywords. The word "ChatGPT" stands out among these, with 56 occurrences. In the sequence, it was observed the highlight for "artificial intelligence" and "education" in a total of 40 and 20 occurrences respectively. In the fourth position stands out the word chatbot with 13 occurrences. There is also ethics, large language models,

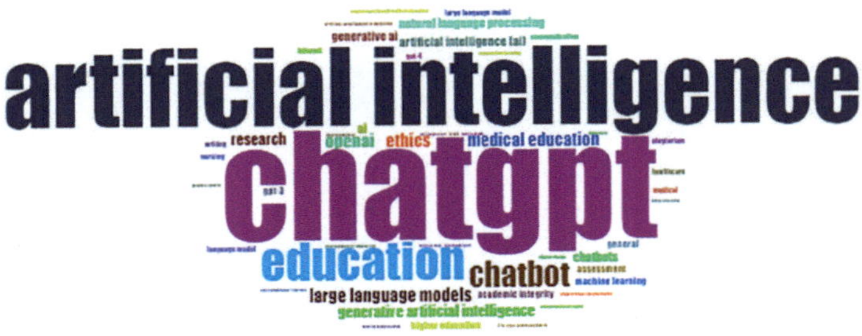

Fig. 7 Tag cloud

Table 3 Most cited documents in the Web of Science database

Author/scientific source	DOI	No. Citation WOS
O'Connor S, 2023, Nurse Educ Pract	10.1016/j.nepr.2022.103537	56
Dwivedi YK, 2023, Int J Inf Manage	10.1016/j.ijinfomgt.2023.102642	22
Cascella M, 2023, J Med Syst	10.1007/s10916-023-01925-4	12
Sallam M, 2023, Healthcare	10.3390/healthcare11060887	12
Alkaissi H, 2023, Cureus J Med Sci	10.7759/cureus.35179	10
Kasneci E, 2023, Learn Individ Differ	10.1016/j.lindif.2023.102274	10
Khan RA, 2023, Pak J Med Sci	10.12669/pjms.39.2.7653	9
Arif TB, 2023, Med Educ Online	10.1080/10872981.2023.2181052	8
Tlili A, 2023, Smart Learn Env	10.1186/s40561-023-00237-x	7
Thurzo A, 2023, Educ Sci	10.3390/educsci13020150	7

medical education, openai with 7 occurrences standing out in fifth place. To analyze these 271 varied words used in 104 documents, a tag cloud was created, shown in Fig. 7, from the retrieved papers, highlighting the key words for this study translated into Portuguese.

5.4 *Analysis of Articles*

The fifth phase, where analytical synthesis of the results is made, the 10 most cited documents globally were listed to answer the research questions: What is the future of education with the use of ChatGPT? The 10 articles described in Table 3 were analysed. It was performed the reading and analysis of the 10 articles described in Table 3.

Of the 10 articles that comprised the corpus of analysis, six did not respond to the research, as they dealt with specific themes related to health, medical education, and clinical management. After reading and analysis of the articles by the researchers, it was identified that 4 articles responded to the main problem of this research which will be explained below.

6 What Is the Future of Education with the Use of ChatGPT

According to him, it is the debut publication by Dwivedi et al. (2023). The advantages, disadvantages, and policy implications of generative conversational AI are evident from interdisciplinary viewpoints, and it is clear that this technology should be used in the future while taking the following considerations into account: One must pay attention to distinctiveness, be aware of different viewpoints, be open to novelty, and be present-focused. Teachers could use this theoretically formed framework to encourage kids to experiment with technology tools. Additionally, this technology can be employed in previously unthinkable ways through teacher and student collaboration on study into its uses and limitations. This study suggests that it's essential to educate kids about the ethical implications of technology use as well as their options.

The second paper by the author was published in Sallam, M. (2023) and is titled ChatGPT utility in healthcare education, research, and practice: a systematic review on the potential views and legitimate issues. Describes how the arrival of ChatGPT may herald the demise of essays as academic writings. It appears that ChatGPT has a lot of potential to transform the field of medical education. It is essential to reassess and enhance the current assessment methods used in health education because ChatGPT may pass renowned exams (like the USMLE) and the potential for abuse that could result in academic dishonesty. The possible impacts of ChatGPT on the efficacy and standard of teaching resources and assessment tools in the field of health education need to be further investigated. Additional studies should be conducted to examine the efficiency of ChatGPT in assisting healthcare students in the development of communication skills, as well as the usage of LLMs in enhancing the accomplishment of intended learning outcomes through personalized and quick feedback to students.

Is "ChatGPT for Good," the third essay Kasneci et al. (2023) authored for his paper, any good? In On Opportunities and Challenges of Large Language Models for Education, it is said that the use of large language models in education for learning has been identified as a potential area of interest due to the diversity of applications they give. All educational levels, including primary, secondary, higher education, and professional development, may benefit from these methodologies for teaching and learning improvement. Because each person has unique learning preferences, abilities, and requirements, great language models also offer a rare opportunity to provide personalized and successful learning experiences. To create individualized learning opportunities for their students, teachers can make excellent use of language models. These models are able to assess students' writing and responses, provide tailored feedback, and suggest resources in line with their particular learning needs. By leveraging this support to help them save time and effort on providing customized materials and feedback, teachers may concentrate on other aspects of teaching, such as creating engaging and interactive sessions.

Participants in the fourth study by Tlili et al. (2023), What if the devil is my guardian angel: ChatGPT as a case study of using chatbots in education, concurred that ChatGPT was effective in providing a complete grasp of diverse (difficult) themes in a straightforward manner. Therefore, it can be argued that ChatGPT will cause a paradigm change in the ways that education is currently delivered and that it will inspire future learning reform that is full of exciting digital possibilities. Students that use ChatGPT's model responses may be able to better understand a number of academic fields. Additionally, by providing an outline of the material that should be covered in their curriculum, ChatGPT can be a useful tool for teachers and instructors. This is crucial for effective instruction design.

7 Final Considerations

The primary goal of this research was to use ChatGPT to map the future of education in light of the bibliometric analysis in the Web of Science database. There were 104 articles in total that were found and fit the article's theme. A bibliometric analysis and a thematic analysis of the papers were created with this goal in mind.

It was discovered that a single publication published in 2022 sparked scholars' interest in using ChatGPT in teaching. There were 103 publications in the region in 2023. With an average of 23% of all publications, or 115 articles, the United States stands out as the country with the most publications in the field. With 45 papers, or 9% of all publications, India is in second place.

However, due to the extensive network of connections that have been built between nations as a result of citations between papers published in the Web of Science database, the following continents have indicated interest in the subject: America, Africa, Asia, Europe, Oceania, and Antarctica.

The six reference authors in the discipline who each have two publications are Siobhan O'Connor, Uli Sauerland, Marlene Agerstrand, Matthias C. Rilig, Malik Sallam, and Kenneth Arne Pettersen Gould.

Finally, in order to respond to the queries Describe ChatGPT. What does the use of ChatGPT mean for the future of education? A sophisticated language model called ChatGPT uses deep learning methods to provide answers to inputs in natural language that are human-like. Deep learning techniques are used by ChatGPT, a complex language model, to deliver responses to inputs in natural language that are human-like. Since 2023, ChatGPT, a member of OpenAI's generative pre-training transformer (GPT) model family, has become more well-known and significant, solidifying its position as one of the most significant language models that are publicly accessible. Since they concentrated on specific problems with clinical management, medical education, and health, six of the ten articles that made up the corpus of analysis did not address the research.

After reading and analyzing several papers, the researchers discovered four that specifically address the main topic of this study. In response to the second question: ChatGPT may be argued to bring about a paradigm shift in the manner in which education is currently provided as well as to serve as an inspiration for a future learning reform that is rich in digital potential. Students who use ChatGPT's model answers may be better equipped to comprehend a variety of academic disciplines. Additionally, by giving an overview – an important component of instructional design – ChatGPT can be a helpful tool for teachers and instructors to remind them of the knowledge and skills that should be addressed in their program.

In the context of higher education, it aims to map how ChatGPT can be used as an instrument of sustainable technology innovation with the application of sustainable development goals.

References

Alqahtani, T., Badreldin, H. A., Alrashed, M., Alshaya, A. I., Alghamdi, S. S., bin Saleh, K., et al. (2023). The emergent role of artificial intelligence, natural learning processing, and large language models in higher education and research. *Research in Social and Administrative Pharmacy, 19*(8), 1236.

Aria, M., & Cuccurullo, C. (2017). Bibliometrix: An R-tool for comprehensive science mapping analysis. *Journal of Informetrics, 11*(4), 959–975. https://doi.org/10.1016/j.joi.2017.08.007

Barney, J., Wright, M., & Ketchen, D. J., Jr. (2001). The resource-based view of the firm: Ten years after 1991. *Journal of Management, 27*(6), 625–641.

Bibliometrix – Home. (2022). Retrieved June 26, 2023, de Bibliometrix.org website: https://www.bibliometrix.org/home/

Dwivedi, Y. K., Kshetri, N., Hughes, L., Slade, E. L., Jeyaraj, A., Kar, A. K., et al. (2023). "So what if ChatGPT wrote it?" multidisciplinary perspectives on opportunities, challenges and implications of generative conversational AI for research, practice and policy. *International Journal of Information Management, 71*, 102642.

Eager, B., & Brunton, R. (2023). Prompting higher education towards AI-augmented teaching and learning practice. *Journal of University Teaching & Learning Practice, 20*(5), 02.

Ghosh, S., Hughes, M., Hughes, P., & Hodgkinson, I. (2021). Corporate digital entrepreneurship: Leveraging industrial internet of things and emerging technologies. *Digital Entrepreneurship, 183*.

Hong, W. C. H. (2023). The impact of ChatGPT on foreign language teaching and learning: Opportunities in education and research. *Journal of Educational Technology and Innovation, 5*(1).

Jalil, S., Rafi, S., LaToza, T. D., Moran, K., & Lam, W. (2023, April). ChatGPT and software testing education: Promises & perils. In *2023 IEEE International Conference on Software Testing, Verification and Validation Workshops (ICSTW)* (pp. 4130–4137). IEEE.

Kasneci, E., Seßler, K., Küchemann, S., Bannert, M., Dementieva, D., Fischer, F., et al. (2023). ChatGPT for good? On opportunities and challenges of large language models for education. *Learning and Individual Differences, 103*, 102274.

Lee, H. (2023). *The rise of ChatGPT: Exploring its potential in medical education*. Anatomical Sciences Education.

Lyu, Q., Tan, J., Zapadka, M. E., Ponnatapura, J., Niu, C., Myers, K. J., et al. (2023). Translating radiology reports into plain language using ChatGPT and GPT-4 with prompt learning: Results, limitations, and potential. *Visual Computing for Industry, Biomedicine, and Art, 6*(1), 9.

Megahed, F. M., Chen, Y. J., Ferris, J. A., Knoth, S., & Jones-Farmer, L. A. (2023). How generative ai models such as ChatGPT can be (mis) used in spc practice, education, and research? An exploratory study. *Quality Engineering*, 1–29.

Nikolic, S., Daniel, S., Haque, R., Belkina, M., Hassan, G. M., Grundy, S., et al. (2023). ChatGPT versus engineering education assessment: A multidisciplinary and multi-institutional benchmarking and analysis of this generative artificial intelligence tool to investigate assessment integrity. *European Journal of Engineering Education, 1-56*.

Peres, R., Schreier, M., Schweidel, D., & Sorescu, A. (2023). On ChatGPT and beyond: How generative artificial intelligence may affect research, teaching, and practice. *International Journal of Research in Marketing*.

Qadir, J. (2023, May). Engineering education in the era of ChatGPT: Promise and pitfalls of generative AI for education. In *2023 IEEE Global Engineering Education Conference (EDUCON)* (pp. 1–9). IEEE.

Rennings, K. (2000). Redefining innovation – Eco-innovation research and the contribution from ecological economics. *Ecological Economics, 32*(2), 319–332.

Rospigliosi, P. A. (2023). Artificial intelligence in teaching and learning: What questions should we ask of ChatGPT? *Interactive Learning Environments, 31*(1), 1–3.

Sallam, M. (2023, March). ChatGPT utility in healthcare education, research, and practice: Systematic review on the promising perspectives and valid concerns. In *Healthcare* (Vol. 11, p. 887). MDPI.

Short, C. E., & Short, J. C. (2023). The artificially intelligent entrepreneur: ChatGPT, prompt engineering, and entrepreneurial rhetoric creation. *Journal of Business Venturing Insights, 19*, e00388.

Strzelecki, A. (2023). To use or not to use ChatGPT in higher education? A study of students' acceptance and use of technology. *Interactive Learning Environments*, 1–14.

Torraco, R. J. (2016). Writing integrative literature reviews: Using the past and present to explore the future. *Human Resource Development Review, 15*(4), 404–428.

Teubner, T., Flath, C. M., Weinhardt, C., van der Aalst, W., & Hinz, O. (2023). Welcome to the era of ChatGPT et al.: The prospects of large language models. *Business & Information Systems Engineering, 65*(2), 95–101.

Tlili, A., Shehata, B., Adarkwah, M. A., Bozkurt, A., Hickey, D. T., Huang, R., & Agyemang, B. (2023). What if the devil is my guardian angel: ChatGPT as a case study of using chatbots in education. *Smart Learning Environments, 10*(1), 15.

Tsai, M. L., Ong, C. W., & Chen, C. L. (2023). Exploring the use of large language models (LLMs) in chemical engineering education: Building core course problem models with chatGPT. *Education for Chemical Engineers, 44*, 71–95.

Wang, F. Y., Miao, Q., Li, X., Wang, X., & Lin, Y. (2023). What does ChatGPT say: The DAO from algorithmic intelligence to linguistic intelligence. *IEEE/CAA Journal of Automatica Sinica, 10*(3), 575–579.

Yang, J., Li, H. B., & Wei, D. (2023). The impact of ChatGPT and LLMs on medical imaging stakeholders: Perspectives and use cases. *Meta-Radiology, 100007.*

Zhong, J., et al. (2018a). A blockchain model for word-learning systems. In *2018 5th International Conference on Behavioral, Economic, and Socio-Cultural Computing (BESC)* (pp. 130–131). IEEE.

Zhong, J., Xie, H., Zou, D., & Chui, D. K. (2018b, November). A blockchain model for word-learning systems. In *2018 5th international conference on Behavioral, Economic, and Socio-Cultural Computing (BESC)* (pp. 130–131). IEEE.

Neolearning as a Methodology That Advances from Experiential and Expansive Learning Theories to Practice

Graziela Grando Bresolin and Patricia de Sá Freire

1 Introduction

One of the most challenging transformations is Digital Transformation. It began as a result of widespread Internet use, and later evolved to the offer of digital technologies; more recently, it is marked by the sharp increase in the offer of connectivity and interaction with mobile devices and smart technologies (Aires, 2020; Pacheco et al., 2020).

Digital transformation is a phenomenon of disruption caused by digital technologies and can influence all areas of activity and human life, leading to structural changes in the means of production and services (Pacheco et al., 2020). It would be no different with education, which is part of this digital and social context that is highly changeable, requiring the ability to anticipate and adapt to change (Riquelme & Balladares, 2022).

There are several factors that are contributing to a sense of urgency in education as a whole to transform itself and prepare for the future. Some of these factors include the technological advances for the development of innovative pedagogies and learning theories, the ever-increasing volume of information due to the growing use of the Internet and new digital technologies, demands of lifelong learning, shift toward technological and open educational resources, and the new generation of learners (Ally, 2019).

For this reason, higher education needs to renew its practices and curricula with the aim of aligning student development with the competencies that are being demanded by this new context, both by society in transformation and by the job market. In addition, current trends in education point to the importance of practical experience in developing these competencies (Ahuett-Garza et al., 2022).

G. G. Bresolin (✉) · P. de Sá Freire
Federal University of Santa Catarina, Florianópolis, Brazil

A. de Bem Machado et al. (eds.), *Digital Transformation in Higher Education Institutions*, EAI/Springer Innovations in Communication and Computing,
https://doi.org/10.1007/978-3-031-52296-3_10

Consequently, the impacts of the transformations currently taking place require teachers to make radical changes in the form and means with which they plan and teach educational actions so that they manage to consider technological, methodological, and generational changes. Teachers can gather relevant information for their practice when they are aware of the characteristics, learning styles, needs as well as expectations for individual and social development of students of the generations born in this period of transformation.

These new paths should allow changes to the teaching model to update class plans, contents, strategies, methodologies, resources, assessments, learning environments, teachers' and students' roles and, above all, the system of relations between these elements. They should also enable other actors to be inserted into the learning network to promote a collaborative, dialog-oriented environment, fostering the co-creation of learning and the improvement development of the competencies expected in the job market (Freire & Bresolin, 2020).

In this direction, Neolearning offers a methodology to support innovation in class. Promotes the application of experiential and expansive learning through the development of Neolearning cycle, with the aim of facilitating the application of learning to social and work practice, which is essential to development of the new competencies. In this sense, it creates a learning environment in which students have the opportunity to experience content, apply knowledge, and interact with the academic and non-academic experts (Bresolin & de Freire, 2023).

In this context, this chapter aims to present Neolearning as a methodology that advances from experiential and expansive learning theories to practice. In pursuit of this aim, a qualitative with exploratory and descriptive approach was selected, using the theoretical foundation of Neolearning to analyze multiple case studies.

2 Experiential Learning

According to the perspective of experiential learning, students have the capacity to learn from their experiences, and these experiences serve as the foundation for knowledge creation. For this reason, experiential learning is "the process whereby knowledge is created through the transformation of experience. This definition emphasizes […] that knowledge is a transformation process, being continuously created and recreated […]" (Kolb, 1984, p. 38).

Notably, this definition enhances the importance of learners' experience by stressing that they have already lived so many things in their life and can, therefore, bring several cases and examples to the classroom if given the chance to be listened to. As a result, this experience, when shared and discussed within educational action, becomes the primary catalyst for learning, as knowledge is constructed through reflective transformation on the experience itself (Baker et al., 2002).

This theory has been applied in educational programs, from individual classes to training, development, and undergraduate and graduate education courses and programs, for the purpose of planning school and university curricula, including the

National Curriculum in New Zealand (Ministry of Education, New Zealand, 2004) and in Singapore (Ministry of Education, Singapore, 2015).

Experiential learning theory was formulated to provide a theoretical and practical basis for educational innovation. This theory was created by David Kolb in 1984 and based on six assumptions:

1. The conceptualization of learning should emphasize it as a dynamic process rather than focusing solely on outcomes. It presupposes that ideas are not rigid and unchangeable components but are shaped and reshaped through experiences. Therefore, students should be involved in a process that fosters learning development, incorporating feedback on the efficacy of their learning endeavors to ensure the continual reconstruction of their experiences.
2. Every instance of learning involves a form of relearning. Learning is enhanced when students have opportunities to clearly talk about their beliefs, ideas, and previous experiences regarding the contents being addressed, allowing for examination, testing, and integration with new and refined ideas.
3. Learning necessitates the resolution of conflicts arising from dialectically opposing modes of adaptation to the world. Conflicts, differences, and disagreements are catalysts for the learning process and should not be avoided; instead, they should be permitted to unfold. In this process, individuals navigate between opposing modules of reflection and action, as well as between feelings and thoughts.
4. Learning is a comprehensive process of adapting to the world. Beyond mere cognitive outcomes, learning fully engages individuals in thinking, feeling, perceiving, and behaving actions.
5. Learning emerges from synergistic interactions between individuals and the environment. It occurs through the balance of the dialectical processes of assimilation of new knowledge, through the experience in existing concepts and the accommodation of new concepts.
6. Learning represents the ongoing process of knowledge creation, based on the constructivist theory of learning, in which knowledge is created and recreated from learners' prior knowledge.

This is why experiential learning is also considered as a dynamic, holistic, and multidimensional model of learner's development, based on the process of learning by doing. Consequently, its applicability extends beyond the confines of formal education classrooms to encompass all facets of life. It operates across various societal levels, impacting individuals' personal lives, relationships, group dynamics, organizational settings, and within the broader context of learning networks (Kolb & Kolb, 2017).

It is driven by the double dialectics of action and reflection, experience and abstraction. The creation of new knowledge results from the combination of ownership and transformation of experience. Experiential learning theory proposes 4 modules for double dialectics to occur two distinct modules for experience appropriation: concrete experience (apprehension) and abstract conceptualization (understanding). Two distinct modules of experience transformation: reflective observation

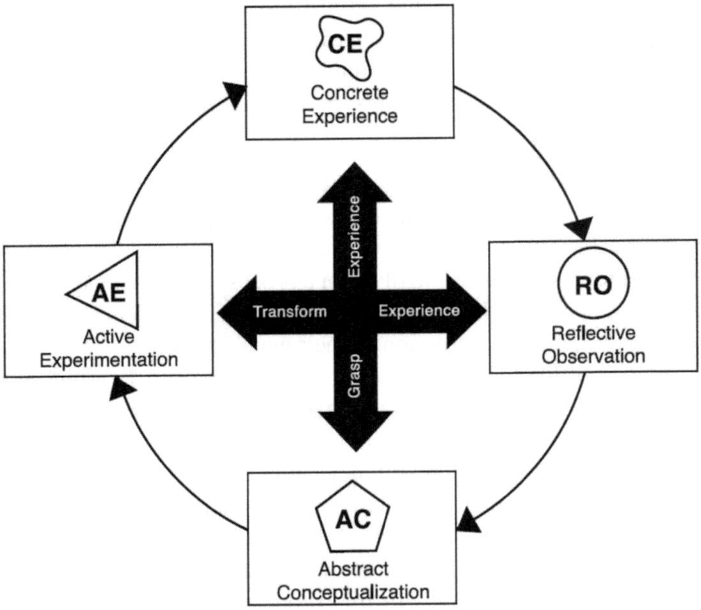

Fig. 1 Experiential learning cycle. (Source: Kolb & Kolb, 2017, p. 11)

(intention) and active experimentation (extension) (Baker et al., 2002; Kolb & Kolb, 2013).

The modules were developed by Kolb (1984) to offer teachers the experiential learning cycle (Fig. 1). Concrete experience (CE) is the first module is related to the realization and explanation of the experience lived by learners throughout life, to prepare them for assimilation and accommodation on the subject to be addressed. The experiences of individuals are valuable in creating meaning for learning and referenced in existing and previously learned mental knowledge and processes. Reflective observation (RO) motivates students to observe and reflect on their knowledge (through introspection) to identify elements; make associations; find connections between the noticeable facts of experience; determine characteristics; realize difficulties and possible choices; express opinions on a particular subject, achieving a level of perception of the need for new knowledge.

Abstract conceptualization (AC) is the third module of the experiential learning cycle, established through activities that drive conceptual interpretation and understanding of concepts, theories, and analysis of the experience, in order to form abstract and generalizable concepts regarding elements and characteristics of the experience. Active experimentation (AE) is the fourth module of the experiential learning cycle, which is action-related and allows the creation of meaning for what has already been understood by individuals. It is a movement toward the outside, for application of the new concepts in a simulated experience, thus promoting the accommodation of learning.

The cycle begins when learners become personally involved in a specific experience (concrete experience). They reflect on the experience that they lived from different points of view and try to understand what it means (reflective observation). Next, they seek, in theory, logical conclusions about the experience lived (abstract conceptualization) and can relate their new conclusions to other theories. These new conclusions underpin decisions and actions in future practice (active experimentation) and generate new concrete experiences.

In addition to the four modules, the experiential learning cycle consists of four types of knowledge (Kolb, 1984), which are developed throughout the cycle. Divergent knowledge makes it possible to generate new ideas and to combine theory with practice. The assimilation of knowledge involves deduction to solve problems, utilizing the logic of an idea to construct theoretical models, organizing concepts, and converting them into coherent explanations. Convergent knowledge is contrary to divergent knowledge, as it enables individuals to make quick decisions in a situation when there is only one correct answer or solution to the problem, and it represents the skills in practical application of ideas. Accommodating knowledge uses creativity to solve problems and develops competence to carry out activities and execute plans.

However, learning only takes place if all modules in the cycle are covered. As emphasis is placed on experience as a driver of learning, these experiences must be present in all four modules, and not only in concrete experience and active experimentation. In other words, all modules in the cycle should promote experiences of thinking, knowing, reflecting, questioning, doing, and learning (Kolb & Kolb, 2018).

For educators, experiential learning consists of the distinctive relationship between the teacher, the student, and the subject under study. The experiential approach places the subject to be learned at the center of learning, to be experienced by both the teacher and the students (Kolb & Kolb, 2017).

It is important to mention that Jarvis (2013, p. 34) criticizes experiential learning theory when he argues: "as a sociologist, he recognizes that all psychological models of learning were flawed, including Kolb's well-known learning cycle, in the sense that they omitted the social and interaction." In the author's view, the constitution of the cycle is reductionist and simplifies the reality of the learning process, which needs to be transposed to other contexts, beyond the classroom. For learning to be processed at a social and institutional level, it needs to be analyzed under other theories that enable new perspectives and the application of new knowledge in work practice.

3 Expansive Learning

For the expansive learning theory t the concept of learning as an activity system that goes beyond the classroom environment and expands into organizations, becoming a social practice. This theory proposes breaking the encapsulation of learning, expanding the object of learning to incorporate the relationships between the educational institution's context and the practical application context. Transforming learning into practice (Engeström, 2002). In addition to the development of expansive learning, the author conducts studies on the learning process through work activities in organizations, with application in the areas of health and education (Engeström, 2013).

Engeström's proposal is based on the previous ones, but it is set up in a third generation of activity theory, as it demonstrates not only the social construction of meaning and learning, but the transformation that occurs in the elements of the system and in the system itself, based on confrontation (divergence) and the creation of opportunities for transformation (convergence). In this understanding, activity gains new structures and objectives: it is focused on solving new problems of the system itself.

For Engeström (2002, p. 179), learning needs to promote the creation and application of new concepts and theories into practice, because learners "[…] have to learn something that is not there, they get their future activity as they create it." From this perspective, learning needs to be designed for the application in the system of activities, that is, for future practice in the social and work environment.

In summary, the objective of expansive learning is to transform an object, product or service and, as a consequence, learning brings about changes in the subjects who develop such activity, modifying their behavior to do their practice, as well as to enhance individuals' cognition. Thus, expansive learning manifests itself primarily alterations in the object of collective activity, involving the transformation of all components within the system of activities. This matrix (Table 1) shows the analysis of five foundations that are interrelated with four questions.

To design the Expansive Learning Matrix, first answer at least four central questions: (1) Who are the learning subjects, how are they defined and located? (2) Why do they learn, what drives them into making the effort required? (3) What do they learn, what are the contents and learning outcomes? (4) How do they learn, what are the fundamental actions of the learning processes?

The fundamentals that support the implementation of the expansive learning theory are presented, according to Engeström and Sannino (2010):

Table 1 Matrix for analysis of expansive learning

	Activity system as unit of analysis	Multi-voicedness	Historicity	Contradictions	Expansive cycles
Who are learning?					
Why do they learn?					
What do they learn?					
How do they learn?					

Source: Engeström and Sannino (2010)

1. Collective activity system is mediated by artifacts and oriented to objects and understood on the basis of the relationships that take place between the network and the other activity systems. The fundamental unit of analysis is the relationships between these two activity systems. Individual and collective actions are considered as independent units of analysis because they perform and reproduce activities, generating actions and operations.
2. Multiple voices compose the systems of activities and interact with each other. In other words, an activity system is always made up of a community of views, traditions, stories, experiences, and multiple interests. This interaction between history and different experiences and cultures is a source of problems, but also a source of innovation and change. Innovation among the systems of activities can occur as long as there are actions that allow two-way dialog and negotiation between interests.
3. Historicity allows the system of activities to be shaped and transformed. "Their problems and potentials can only be understood against their own history" (p. 136, our translation). Thus, the very history of the individuals that compose the system needs to be studied and understood, as well as the history of activities and objectives, ideas, and theoretical tools.
4. Contradictions serve as sources of change and development. It's important to note that contradictions differ from problems or conflicts; they arise when creative and structural tensions accumulate within and between activity systems. Because activity systems are typically open to the external world, contradictions can arise when a new external element is adopted to compose the system. This new element can collide with another existing element in the system, generating tensions that can result in innovations or new activities.
5. Expansive transformations generate long cycles for the development of activity systems. When serious contradictions are found in an activity system, some individual participants begin to question and deviate from the norms established by the system which they are a part of.

In certain instances, this can evolve into a collaborative vision and a collective endeavor for change. An expansive transformation takes place when the purpose and rationale of an activity are reconceptualized to provide an understanding of a broader range of possibilities than in the previous mode of the activity.

Expansive learning entails the creation of a new expanded object and an object-oriented activity pattern. This process includes the development of a theoretical concept for the new activity, built upon the understanding and modeling of the initial simple process. This, in turn, gives rise to the new activity and produces its diverse concrete manifestations (Engeström & Sannino, 2010).

Engeström and Sannino (2010) posit that the transition from the abstract to the concrete is accomplished through specific epistemic or learning actions. These actions collectively constitute the expansive learning cycle, as illustrated in Fig. 2 below.

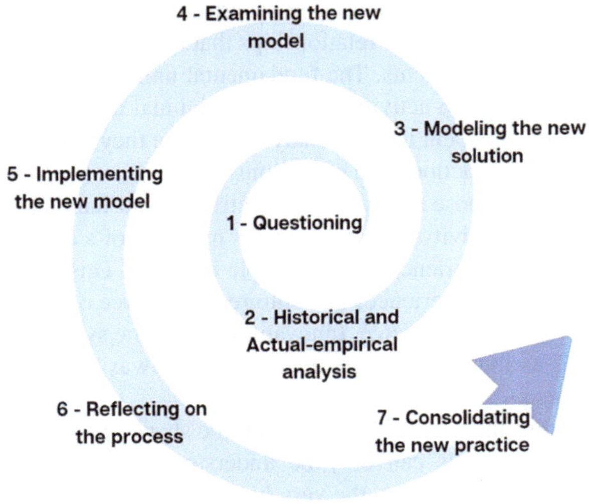

Fig. 2 Expansive learning cycle. (Source: Created by the authors, based on Engeström & Sannino, 2010)

Seven stages of the expansive learning cycle are described below:

1. Questioning, criticizing, or rejecting some aspects of existing practice or knowledge is the first stage of the expansive learning cycle, which is also referred to as questioning learning.
2. Historical and actual-empirical analysis involves engaging in the mental, discursive, or practical transformation of a given situation to uncover causes or mechanisms that can elucidate it. The analysis triggers the question "Why?" and explanatory principles. There are two types of analysis: the genetic-historical one, which aims to elucidate the situation by tracing its origins and evolution, and the real empirical one, which attempts to explain the situation by identifying causes and effects.
3. Modeling the new solution by obtaining observable and transferable information and knowledge, which means building an explicit and simplified model of the new idea that explains and provides a solution to the problem or problematic situation that is actually found.
4. Examining the new model, run it, operate it, try it, and test it with the objective of fully understanding its dynamics, potentials, and limitations, for the purpose of continuous improvement.
5. Implementing the new model through practical applications to further develop the previously modeled concept in a practical application, allowing observation and analysis of its action and execution.
6. Reflecting on the process so far, from steps 1 to 6, to identify needs for improvement or to move on to the last action.
7. Consolidating the new practice to make it stable and a part of the new system.

The expansive cycle was initially applied and designed for long-term large-scale projects, but several shorter cycles of learning actions were found to occur within this cycle, enabling the development of all phases of the larger cycle. The expansive learning cycle begins to be developed when a relatively stable and existing activity pattern begins to be questioned. Correspondingly, the cycle ends when a new activity pattern becomes consolidated and stable, again (Engeström & Sannino, 2010).

4 Neolearning as an Advancement of Experiential and Expansive Learning Theories

Neolearning establishes a theoretical and practical foundations for the application of the theories of experiential learning (Kolb, 1984; Kolb & Kolb, 2017) and expansive learning (Engeström, 1987; Engeström & Sannino, 2010). These theories consider the adults' prior knowledge and incorporating their experiences into the learning environment while expanding the learning object to encompass the learning network (Engeström, 1987; Engeström & Sannino, 2010; Kolb, 1984; Kolb & Kolb, 2017).

The experiential learning theory serves as the foundational framework for Neolearning, guiding the process of learning through experience. It identifies learners' diverse learning styles and recognizes the varied roles played by teachers. According to Kolb (1984), individuals have the capacity to learn from their experiences, and these experiences serve as the foundation for knowledge creation. Neolearning embraced principles, guidelines, and frameworks of the experiential learning cycle to develop its methodology, but it was eventually found that "education is not preparation for life; education is life itself" (Dewey, 1978). In other words, learning takes place beyond educational action, because learning exists before, during, and after the period of education, and each of the learnings acquired is equally important for learner development.

For this very reason, universities, academic or corporate, embrace the experiential learning process and advance beyond the classroom context, breaking the protective membrane that traps teachers and learners in a disconnected world of practice. To achieve this level of learning, Neolearning recognizes the foundations of the expansive learning theory, because it allows us to overcome the environment of educational action to apply new knowledge in other learning contexts.

The expansive learning theory significantly contributes to Neolearning by fostering an enhanced acceptance of the "learning object" that incorporates the relationships between the higher education context and the context of practical application. This acknowledgment is crucial because learning extends beyond the confines of the traditional classroom environment and expands into broader learning networks It is no longer considered as a fixed content of the activity (as it is carried out today); emphasis lies in how this content relates to the activity, based on practical application (in a context transformation perspective) (Engeström & Sannino, 2010).

Under these conditions, it can be argued that the teaching and learning process should "climb over" the university walls and turn learning into practice. In essence, the learning process should transcend the boundaries of the traditional classroom setting and extend into expansive learning networks. Consequently, Neolearning adopts this expansive learning approach to facilitate networked learning. The objective is to promote co-production and address complex problem-solving projects collaboratively with network stakeholders. These stakeholders include both internal collaborators, fostering interaction between units or different areas and departments, and external partners, such as other universities, organizations, or society at large.

This approach aims to create a dynamic and interconnected learning environment that reflects the complexity of real-world challenges. The expansion of the learning object to a collective system of activity implies a transformation in the teaching activity in universities, resulting in the design of a collective instrument of interaction and learning between learners, teachers, community, organizations, and among the actors within the system.

In other words, two systems are now being considered, rather than just one. In addition to the system composed by the teacher, the student, and the object of the study, there is also the system of future application of learning. Thus, the learning process should be reshaped, in order to contemplate two systems of activities in constant interaction, which enable sharing and co-production of the learning object.

A total of 6 objectives have been set in order for the principles of Neolearning to be faithfully followed:

1. Respect different learning and teaching styles; shift the focus of teaching towards the knowledge to be taught and to students' learning process;
2. Establish a co-production process of new knowledge in learning networks;
3. Allow the various actors of education to alternate between teaching and learning roles;
4. Create collaborative environments to bridge the gap between learners and teachers;
5. Use agile methods and learning assets to develop competencies;
6. Develop expandability and readiness to transfer learning to practice.

4.1 Neolearning Cycle

The andragogical and heutagogical approaches and the theories of experiential learning and expansive learning provide the theoretical foundations for the different moments in a class known as the Neolearning cycle (Fig. 3). This cycle comprises five interconnected, dynamic, and fluid modules aimed at fostering the development of technical, digital, and socio-emotional competencies. The five modules of the Neolearning cycle are designated: retrieve, reflect, discover, test, and apply.

The first module promote simulated experiential activities recover the learner's previous knowledge and experiences and encourages them to realize what they already know about the object to be studied. The second module includes an open

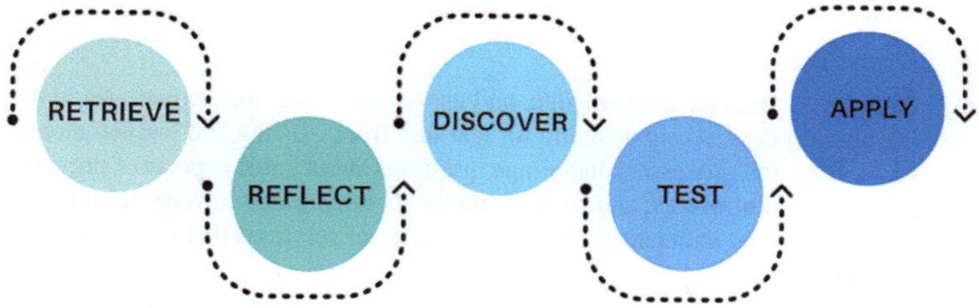

Fig. 3 Neolearning cycle. (Source: The authors)

space for reflection, observation, discussion, and analysis on the experience lived, without criticism and without determination of right and wrong, provides the freedom to identify the gaps to be filled and the motivation to discover something new. The third module addresses the understanding concepts, theories, and models that underlie the background for generalizing of knowledge to be applied in different experiences (Bresolin & de Freire, 2023).

The fourth module included the test of the experiential activities for practical application of knowledge making it possible to test new ideas and concepts in situations closer to reality. The fifth and last module covers the application of supervised or monitored experiential practices to help learners go beyond and overcome the limitations of the controlled environment and apply learning to social and work practices (Bresolin & de Freire, 2023). The description of the application guidelines for each module of the Neolearning cycle is presented in Table 2.

These guidelines support the planning and execution of classes, courses, and education programs aligned with Neolearning to promote competencies. By adhering to these guidelines, teachers can create learning environments that actively engage students in processes such as discovery, reflection, analysis, collaboration, and co-production. Additionally, these guidelines emphasize the importance of fostering a culture of knowledge and experience sharing, as well as providing opportunities for testing and practical application of knowledge.

A key consideration in the application of these five modules in classroom enable the recognition of learners' prior knowledge and motivation to learn, apply, and transfer knowledge. Moreover, contribute to the creation of a dynamic and inclusive educational ecosystem that nurtures continuous growth and development among learners.

Neolearning developed on the basis of the dialog between the principles and guidelines of experiential and expansive learning. The focus of experiential learning is the creation of tacit knowledge, based on practice in situations that are more closely connected with reality. For expansive learning, however, learners' tacit knowledge is created when the object of learning is expanded to include the relationship between the context of higher education and that of application to the work practice, through the co-creation of learning networks that transcend institutional boundaries.

Table 2 Application guidelines of the Neolearning cycle

Module	Application guidelines
Retrieve	Appreciate, respect, acknowledge, and use learners' previous experience in the learning process; create meaning for learning from existing and learned mental knowledge and processes; encourage questions about some aspects of previous practice and knowledge and possible applications; promote divergent knowledge so that learners can understand the different types of solutions that are possible for the same problem
Reflect	Provide space for learners to share their experiences and express their thoughts and reflections on the object of study; promote moments of reflection, individually and then as a group, on shared experiences; ask questions, encourage different opinions, generate discussions, and share opinions about events (experienced in the previous module); develop learners' ability to observe their experiences from different perspectives in order to develop a critical reflection on aspects of the context; highlight elements of practice associated with the fundamentals and theories that will be addressed later, without judging right and wrong; motivate learners about what is still to be known
Discover	Request the research of concepts, theories, and models on the subject to be addressed in bibliographic and documentary references; present theoretical foundations; relate the theories to experiences and analysis shared in the previous modules; encourage logical and justified conclusions about the experience; facilitate the development of assimilative knowledge, using deduction to solve problems, making use of the logic of an idea to create theoretical models, organize ideas, and transform them into coherent explanations
Test	Create group activities to apply the concepts, theories, and models learned to case studies, simulations, workshops, or laboratory experiments; promote space that facilitates the development of creativity and the generation of multiple ideas, during the attempts to apply knowledge in the search for solving problems and testing solutions; promote the creation of convergent knowledge, making it possible to find a more appropriate solution to a specific problem, with collective decision-making
Apply	Invite non-academic experts from the area of application of the knowledge dealt with to participate as teachers in this module; create learning networks between learners and collaborators related to the practice environment, breaking the boundaries of the simulated learning environment; expand the learning object to the relationships between the simulated context and the practical application, projecting the application for possible social or professional practices; allow observation, reflection, and evaluation of the process of consolidating new knowledge in practice; encourage the development of new models for the practice experienced, so that new knowledge can be institutionalized; create space for discussion after the practical experience, so that the new knowledge can be accommodated and internalized; ask about the lessons learned and good practices to be shared from the experience

Source: Bresolin and Freire (2023, p. 7), based on Engeström and Sannino (2010), Kolb and Kolb (2017)

4.2 Neolearning Mechanisms

In addition to the guidelines, five mechanisms have been listed to facilitate stakeholder recognition, the establishment partnerships for collaboration, sharing experiences, and integrating knowledge with the network stakeholders to promote the transfer of learning to practice and the co-production of the common good (Bresolin & de Freire, 2023).

The first Neolearning mechanism is called stakeholder recognition. This mechanism supports the development of processes for identifying, recognizing, and classifying stakeholders. Within Neolearning, educational programs are strategically planned and executed to involve a diverse of stakeholders within the network. This inclusive approach enables stakeholders to not only experience and broaden their learning but also to apply knowledge beyond institutional boundaries. Consequently, these engaged stakeholders form collaborative partnerships for the co-production of projects spanning various sectors. Together, they tackle complex issues that align with collective interests and expectations (Bresolin & Freire, 2023).

The second mechanism describe the establishment partnerships for collaboration. This mechanism involved managing the relationship and interaction to motivate collaboration to problem-solving. The aim is to achieve anticipated results while meeting the expectations of all stakeholders involved. In this context, Neolearning cultivates dialogic environments and encourages collaboration among diverse stakeholders to address complex problems. Additionally, it leverages educational and collaborative technologies to facilitate interaction, extend reach, and enhance interconnection between internal and external partners (Bresolin & Freire, 2023).

The third mechanism was named Sharing experience and integrating knowledge with network stakeholders, because Neolearning is established as a platform that integrates internal and external knowledge, facilitating its widespread sharing. This integration of knowledge is a dynamic process through which interactions with the internal and external environment lead to the creation of new insights and understanding. The outcomes extend beyond individual participants to benefit all involved in the value network. Ultimately, this collaborative knowledge integration contributes to the achievement of results that not only serve the participants but also have broader positive implications for the community and society at large (Bresolin & Freire, 2023).

For each module of the Neolearning cycle, it is recommended to employ diverse teaching and learning strategies, leverage educational technologies, and embrace active and agile methods to introduce innovative experimental and expansive practices. These recommendations align with the fourth mechanism, termed the "transfer of learning into practice," as students are more likely to be motivated to apply acquired competencies in their social and professional contexts (Bresolin & Freire, 2023).

From this standpoint, learning transfer involves the effective and continuous application of knowledge, skills, and attitudes acquired, consolidating competent performance in one's work practice. The Neolearning cycle is designed with the specific goal of facilitating the application of learning in practice. Each module incorporates various activities geared toward promoting learning, with the final two modules specifically allowing for application and testing in a simulated or quasi-real environment that mirrors the conditions encountered in the actual working environment. This approach enhances the practical relevance of the acquired competencies, fostering a seamless transition from learning to real-world application (Bresolin & Freire, 2023).

The final mechanism of Neolearning supports the co-production of the common good. The core objective of implementing Neolearning is to foster the co-production of knowledge as a shared asset for all involved stakeholders. This goes beyond the integration of knowledge internally and external to the network participants; it also aims to collaboratively devise solutions to societal challenges. In this way, Neolearning not only enhances the collective knowledge base but also contributes to the joint development of solutions that address broader issues faced by society as a whole (Bresolin & Freire, 2023).

5 Research Methodology

The research is characterized by a qualitative approach (Creswell, 2010), intending to present Neolearning as a methodology that advances experiential and expansive learning theories to practice. This chapter is classified according to Triviños (2015) as exploratory and descriptive. It is exploratory because it describes the theoretical foundations of Experiential Learning Theory, Expansive Learning Theory, and the Neolearning methodology. It is also considered descriptive, because it describes the cycle, principles, guidelines, and the mechanisms to Neolearning. The research also illustrates the application of Neolearning in multiple cases, validating the methodology. In terms of methodological procedures, a narrative literature review was selected on the experiential and expansive learning theories and the Neolearning methodology. The choice of a multiple case study aligns with Yin's (2014) perspective, as it constitutes an empirical research strategy enabling the exploration of one or more cases to investigate the same phenomena. Multiple case studies are deemed to generate more robust evidence than a single case study. To fulfill this objective, three cases were carefully selected.

In the A Case, eight teachers were trained to implement Neolearning, delivering eight lessons as part of a project to 50 learners. Moving on to the B Case, Neolearning was applied in undergraduate disciplines. Finally, in the C Case, Neolearning was utilized for planning an online course, along with training tutors for virtual environment facilitation.

Data collection and analysis were carried out by the authors, actively involved in planning and executing Neolearning applications in the presented cases. The process involved maintaining a field diary to document applications, best practices, and lessons learned. Additionally, assessments were conducted to measure the satisfaction of both teachers and students with the application of Neolearning, encompassing feedback from participants in the form of recorded comments and observations.

6 Multiple Case Study Presentations

This section presents the applications of Neolearning in four different case studies.

6.1 A Case

The inaugural application of Neolearning took place in the A Case, spear-headed by the Federal University of Santa Catarina (UFSC) in collaboration with universities, institutions, and organizations within the innovation ecosystem of the city of Florianópolis, Brazil. This initiative aimed to promote co-production among the different stakeholders of the innovation ecosystem. The focus was on creating human capital and innovative, sustainable enterprises, offering opportunities for the development of technical, digital, and socio-emotional competencies.

In this project, eight teachers were trained to implement the Neolearning cycle across eight lessons, aimed to motivate the development competencies in fifty students. These students were organized into ten groups composed of five students and each group represented an organization. Their task was to develop sustainability solutions and a product that could generate a positive business impact. Throughout the project, rounds of pitch presentations were conducted to continuously refine solutions until the final presentation. The projects with the most innovative solutions were validated by the organizations for future applications and adaptations.

The A Case included the stakeholders from the innovation ecosystem of Florianopolis, including academic participants such as professors from various knowledge domains (engineering, business, psychology, education, technology, and knowledge management) and non-academic, with the involvement of leaders and collaborators from eight companies.

Collaborative partnerships materialized among different teachers, both academic and non-academic, who shared their knowledge to support learners in project development. Furthermore, collaboration with organizations in the innovation ecosystem facilitated project development in tandem with leaders and employees, who contributed internal and external knowledge. The resulting products and services proposed by the groups were not only aligned with organizational expectations but also presented solutions to real-world problems.

In the A Case, the learners were not only able to apply their learning to develop impactful solutions but also transferred their newfound knowledge to future projects, thereby developing competencies valued by organizations. The sustainability solutions and the impact product generated during the project proved beneficial to all the participants that are involved, including the organizations, university, community, and society.

The students said that classes structured according to the five modules proposed by the Neolearning cycle are more motivating than traditional classes, or classes that don't propose these modules. In the results, students also scored positively on the elements used in the classroom, such as use of different technologies, active strategies and methods (Design thinking, case study, project-based learning, collaborative learning, team-based learning), and agile (SCRUM and Sprint, validations along the way for improvement cycles of the final project delivery), diversity of resources used during classes (different physical materials and canvas), practical classes to develop the two challenges with specific classes and content to help each stage of the project elaboration, development of the innovative ideas to solve the challenges of each organization and at the same time generate entrepreneurship and think about the social impact of the solutions.

6.2 B Case

The Neolearning methodology has been applied since 2018 in B Case, specifically in courses focused on entrepreneurship, creativity, and innovation. During the COVID-19 pandemic, these courses have transitioned to online formats. This adaptation facilitated the validation of the Neolearning methodology in both in-person and remote settings. The courses have effectively employed the Neolearning cycle in the innovation process, fostering the development of innovation and entrepreneurship projects that address real-world problems encountered by undergraduates.

The courses aim to cultivate creativity, innovation, and entrepreneurship. They employ various strategies and dynamics designed to stimulate idea generation, foster creativity, and impart essential concepts related to innovation. The Neolearning cycle is seamlessly integrated throughout the duration of the courses. It involves retrieving learners' existing knowledge on key topics, identifying, reflecting, and analyzing contextual problem situations to comprehend the stages of the innovation process. Subsequently, learners are equipped to propose solutions, which are then tested through simulation in the classroom and eventually presented in the form of a product or service pitch.

The application of Neolearning cycle during the course helps the students apply their acquired knowledge to the development of products and services, thereby cultivating competencies for future endeavors. In conjunction with the cycle, the different teaching and learning strategies contributed to the innovation and motivation of the students in class. The variety of strategies including individual and group presentations, flipped classroom approaches, teamwork dynamics, discussions, contextual and situational diagnostics, academic research, syntheses, seminars, creation of prototypes for innovative classes, simulations, proposition of solutions, and presentation of idea pitches.

The involvement of different participants is due to the diversity of students, encompassing various courses and addressing real challenges encountered within the academic community. This approach fosters a collaborative environment, encouraging the co-production of solutions among students from different academic disciplines. Additionally, subject matter experts were enlisted to deliver lectures on topics related to innovation, contributing to the enhancement of the products or services proposed by the students.

6.3 C Case

The Neolearning methodology has also been applied in C Case, specifically within an online course centered on the Sustainable Development Goals (SDGs) outlined in the 2030 agenda for the development of municipal agents. This course endeavors to equip governmental, academic, and civil society representatives from all Brazilian states with the skills to formulate sustainable initiatives and projects for their cities. Beyond supporting city-centric initiatives and projects, the course plays a crucial role in fostering the development of fundamental competencies for the effective municipalization of the 2030 agenda.

Neolearning has been implemented in this context to instill a sense of awareness, mobilization, and sensitivity to sustainability among participants. It facilitates the collective application of knowledge in the practice of municipalities. In addition to formulation of initiatives and co-production projects that contribute to the enhancement of cities. In this particular course, twelve tutors were additionally trained to develop activities and apply the Neolearning methodology in the online learning environment.

In the C Case, the course was designed to encompass city administration, involving different levels of government power and universities, as well as a non-governmental organizations, communities, and societies. The goal was to foster an environment where all these stakeholders could collaboratively generate initiatives and projects to enhance their cities. Partnerships were established between participating city governments and organizations that dedicated the time of their employees to engage in training and meetings for co-producing projects and initiatives. Additionally, universities collaborated to conduct research and planning for different modules of the course.

The Neolearning cycle was applied in the planning and structuring of the course in the virtual learning environment, along with the three stages of environmental education (sensitize, raise awareness, and mobilize). Practical examples, activities, and case studies of the application of initiatives in cities were also used. This helped students understand how to put their knowledge into practice. As a result, the Neolearning methodology applied proved necessary for achieving the course's objectives, helping to develop the competencies of public agents and civil society for the municipalization of the 2030 Agenda.

The projects and initiatives developed by participants facilitate the transfer of learning about the SDGs to organizations, universities, and the city. The application of Neolearning in planning the C Case allowed for the progression through the stages of environmental education, fostering sensitization, awareness, and mobilization among public agents and society. This trans-formative process aimed to turn societal issues and problems into impactful projects and initiatives.

By creating an environment conducive to individual and group learning, Neolearning serves as a motivator and supports the development of sustainability projects for the implementation of the SDGs, rooted in the specific challenges faced by cities. Consequently, initiatives and projects emerge from the co-production efforts of city stakeholders, resulting in solutions that contribute to the common good of citizens.

7 Conclusion

Digital transformations in higher education propose the implementation of innovation and entrepreneurship in the classroom to solve complex problems, as well as the development of competencies aligned with the new demands of the job market. For this to happen, learning needs to relate theory to practice, proposing experiences that motivate learning and make it easier to get closer to the job market.

This scenario requires an interaction between the teaching-learning process and applied content. This interaction is achieved through real and experiential situations of the work environment, with contextualized activities that integrate theory and practice and utilization of new technologies. This way, they can offer an environment for integration between theory and practice, thus preparing students to become citizens with awareness, critical thinking, and readiness to apply their knowledge to solving complex problems posed by current society, which is undergoing digital transformation.

In this sense, Neolearning was developed as a methodology based on experiential and expansive learning theories to create a practical learning environment to facilitate the transfer of learning and the development of competencies.

The distinct modules of the Neolearning cycle (retrieve, reflect, discover, test, and apply) provide opportunities for the student to apply previous knowledge in an experiential activity, and reflect on this application to motivate them to learn something new. This new knowledge will be tested in other experiential activities and applied in practical projects that provide dialog, collaboration, and co-production with academic and non-academic experts, promoting the sharing of experiences and knowledge to solve complex problems.

The innovation of Neolearning is based on practical application from the start of the class, through individual or group reflection, to generate new knowledge that will be applied in a new practical experience and expanded to include interaction with different specialists in the ecosystem.

For future work, we recommend ongoing practical applications across various knowledge domains and continued empirical research to validate the Neolearning cycle, mechanisms, and the competencies developed through the application of the methodology. This iterative process ensures the ongoing refinement and enhancement of Neolearning as a dynamic and effective educational approach. As well as a deeper understanding of the mechanisms and the development of a tool to evaluate the applications.

Acknowledgments This study was financed in part by the Coordenação de Aperfeiçoamento de Pessoal de Nível Superior – Brasil (CAPES) – Finance Code 001.

References

Ahuett-garza, H., et al. (2022). Train the trainers in industry 4.0: A model for the development of competencies in non-synchronous environments. *International Journal on Interactive Design and Manufacturing, 16*, 775–789.

Aires, R. W. d. A. (2020). *Desenvolvimento de Competências Gerais para a Sociedade em Transformação Digital: uma Trilha de Aprendizagem para profissionais do setor industrial.* Mestrado em Engenharia e Gestão do Conhecimento pela Universidade Federal de Santa Catarina.

Ally, M. (2019). Competency profile of the digital and online teacher in future education. *The International Review of Research in Open and Distance Learning, 20*(2). https://doi.org/10.19173/irrodl.v20i2.4206

Baker, A., Jensen, P., & Kolb, D. (2002). *Conversational learning: An approach to knowledge creation*. Quorum.

Bresolin, G. G., & de Freire, P. S. (2023). Neolearning methodology applied to the networked university. *Journal of Information & Knowledge Management, 22*(4), 1–21.

Creswell, J. W. (2010). *Projeto de pesquisa: Métodos qualitativo, quantitativo e misto* (3rd ed.). Artmed.

de Freire, P. S., & Bresolin, G. G. (2020). As novas competências exigidas à tríade educacional para apoiar a formação do novo jovem adulto universitário. *Revista NUPEM, Campo Mourão, 12*(27), 129–143.

dos Pacheco, R. C. S., dos Santos, N., & Wahrhaftig, R. (2020). Transformação digital na Educação Superior: modos e impactos na universidade. *Revista NUPEM, Campo Mourão, 12*(27), 94–128.

Dewey, J. (1978). *Vida e educação. Tradução e estudo preliminar por Anísio Teixeira* (10th ed.). Melhoramentos/Fundação Nacional de Material Escolar.

Engeström, Y. (1987). *Learning by expanding: An activity-theoretical approach to developmental research.* Orienta-Konsultit.

Engeström, Y. (2002). Non scolae sed vitae disimus: Como superar a encapsulação da aprendizagem escolar. In D. Harry (Org.), *Uma introdução a Vygotsky*. Edições Loyola.

Engeström, Y. (2013). Aprendizagem expansiva: por uma conceituação pela teoria da atividade. In K. Illeris (Org.), *Teorias contemporâneas da aprendizagem.* Penso.

Engeström, Y., & Sannino, A. (2010). Studies of expansive learning: Foundations, findings and future challenges. *Educational Research Review, 5*, 1–24.

Jarvis, P. (2013). Aprendendo a ser uma pessoa na sociedade: Aprendendo a ser eu. In: *Teorias contemporâneas da aprendizagem.* Organizador: Knud Illeris.Tradução: Ronaldo Cataldo Costa. Revisão técnica: Francisco Silva Cavalcante Junior. Porto Alegre: Penso.

Kolb, D. A. (1984). *Experiential learning*. Englewood Cliffs.

Kolb, A. Y., & Kolb, D. A. (2013). *The Kolb learning style inventory 4.0: A comprehensive guide to the theory, psychometrics, research on validity and educational applications.* Hay Resources Direct.

Kolb, A. Y., & Kolb, D. A. (2017). Experiential learning theory as a guide for experiential educators in higher education. *A Journal for Engaged Educators, 1*(1), 7–44.

Kolb, A. Y., & Kolb, D. A. (2018). Eight important things to know about the experiential learning cycle. *Australian Educational Leader, 40*(3), 8–14.

Riquelme, G. M., & Balladares, J. (2022). Challenges in curricular innovation in university training from the perspective of academics. *Interciencia, 47*(5), 181–190.

Triviños, A. N. S. (2015). *Introdução à pesquisa em ciências sociais: a pesquisa qualitativa em educação* (1st ed., p. 23). Atlas.

Yin, R. K. (2014). *Case study research: Design and methods* (5th ed.). Sage.

Adopting Blockchain for Educational Qualifications in Italy: The Experience of the University of Turin

Katia Presutti and Francesca Natale

1 Introduction

Blockchain technology is gaining popularity across numerous sectors and universities in Europe and beyond (Raimundo & Rosário, 2021). It is a worldwide, cross-industry, and groundbreaking technology expected to drive growth in the global economy over the next several decades. Europe faces challenges in obtaining educational credentials, including continuous professional development, facilitating non-formal learning recognition, and standardizing the process of issuing and recognizing credentials, especially for open learners and migrants (Calandra et al., 2022). As suggested in an editorial by Sousa et al. (2022), several scholars are examining the role of blockchain in different levels of education. In particular, what they are interested in are the learning outcomes and, in the case of blockchain, the possibility of equal access and widely recognized certifications. Additionally, as stated by Reis-Marques et al. (2021), blockchain can be used to advance the level of protection and communication of digital certificates. Thus, technology has a dual role. First, to protect against fraud. Second, to promote the communication and interchange of certificates between educational institutions and employers. In addition, Raimundo and Rosário (2021), discovered that blockchain can be used not only among students as an award for educational completion but also externally by promoting easy mobility of certificates even for international pathways (e.g., Erasmus). Despite the existing widespread interest, only a few studies analyze case

K. Presutti (✉)
IT Systems, Website Management, E-learning Division, University of Turin, Turin, Italy
e-mail: katia.presutti@unito.it

F. Natale
Innovation and Internationalization Division, University of Turin, Turin, Italy
e-mail: francesca.natale@unito.it

A. de Bem Machado et al. (eds.), *Digital Transformation in Higher Education Institutions*, EAI/Springer Innovations in Communication and Computing,
https://doi.org/10.1007/978-3-031-52296-3_11

studies and real applications. Given this gap, our chapter aims to address the following research question (RQ):

How can blockchain foster educational qualifications in universities?

Considering the gap previously described, our study aims at analyzing the feasibility, challenges, benefits, and risks of implementing blockchain technology in education, focusing on *formal* and *non-formal credentials* for policymakers and non-specialist audiences. The study blends desk research with assessments from innovators, pointing out that the structure of the technology adopted defines the fundamentals and weaknesses of the study itself. This chapter explores the potential of blockchain technology in managing intellectual property, educational grants, and enhancing learners' control over their personal data. More precisely, we aim to (1) provide an introduction to this technology and its social value; (2) identify and address the key issues influencing stakeholders in evaluating the use of blockchain technology within a university system; (3) explore how blockchain technology can enhance academic institutions' ability to protect their brands and reputations while also enabling individuals to enhance their learning portfolio.

In addressing these aims, we use some case studies of Italian universities, focusing on the University of Turin, which adopted the Blockcerts project by applying the public blockchain of Ethereum.

2 Literature Review

2.1 Definitions and Application

Blockchain technology is a secure, shared, and verified infrastructure for learning achievements, particularly certifications. It stores the issuer, the receiver, and the signature of each certificate in a public database on thousands of computers worldwide. It can be verified with certainty that the certificate has been issued and received by the same individuals. The certificate can be verified by anyone with blockchain access using open-source software, eliminating intermediary parties and ensuring validity even if the original organization no longer exists or has access. Blockchain certificates can only be destroyed by deleting every copy on every computer hosting the software. A link to the original document will be created and it will be held only by the user, preserving document privacy.

Issuer: Blockchains provide ideal requirements for issuers to issue certificates, including proof of issue, expiration time settings, revocability, and a secure, minimal ongoing burden.

Recipient: Blockchains provide ideal requirements for a certificate from a recipient's perspective, including independence, ownership, control, verifiability, and permanence. The recipient owns the credential, can prove ownership, curate credentials, and associate credentials with an established profile.

A third-party verifier: To ensure credential validity, a third-party verifier must verify the certificate's authenticity and integrity, ensuring it matches the issuer's original intent and it is not forged, as per standard requirements.

Certifying Identity: A person's identity is the sum of all their personally identifiable information (PII). When someone wishes to confirm their identity to another person or institution, they often share a significant amount of their personally identifiable information.

The blockchain technology offers a new concept of self-sovereign identity, allowing users to store their personal information on a personal device and to share it with third parties only when needed. This digital equivalent of keeping paper certificates at home allows users to certify their identity without sharing the underlying data, reducing the risk of abuse, fraud, and theft. This overcomes traditional methods of storing and sharing personal information, such as handing over data to third parties.

Issuing a Diploma Supplement on a Blockchain: A degree certificate contains minimal information, such as date, awarding institution, awardee, and title of the degree. This information can be stored on a blockchain for the creation of a public database or as a hash of the certificate for digital certificate security. European Higher Education Area graduates also receive a diploma supplement with their qualification, which includes the level, function, contents, results, certification, and details of the national higher education system. This information is not suitable for storage in a ledger and it would be prohibitively expensive to store directly on a blockchain. Certificates can be represented as tokens on a custom blockchain, allowing for easy transfer and storage of information. They can feed a database with private and public information. The advantage of issuing certificates directly on a blockchain is that they become immutable and permanent, rather than just mere proof of signing.

2.2 Technical Characteristics of Blockchain Technology

This paragraph provides an in-depth understanding of the technical foundations of blockchain technology, aiming to provide a better understanding of its capabilities.

Hashing is a short code used to create a unique ID for a digital document. A hash generator allows users to upload a string of text, making it an anti-tampering device. However, hashes are one-way, meaning they cannot be generated from a document mathematically

Blockchain secures each transaction block by including a hash of the information block and of the previous block, ensuring no modifications or tampering by all the parties involved.

Public and private keys – A public key is a publicly available ID number, while a private key is a mathematically linked password. Users can authenticate ownership of a public key by entering their private key details into software, but this function cannot be reversed.

Issuing certification using digital signatures – Digital certification solutions use digital signatures to issue certificates, unlike electronic signatures, which are easily copied or forged and lack verification or standardization. Digital signatures verify a specific document's signature by a specific person. Digital signatures enable certificate issuance by generating unique stamps and preventing tampering. They require a public key and a linked password for each signatory.

Architecture of Blockchain-Secured Digital Certificates Blockchain stores the certificate issuer, the receiver, and the signature in a public database on thousands of computers worldwide, ensuring secure and transparent transactions. Digital credentials, stored on hard drives, mobile wallets, or printed materials, enable users to verify the identity of a certificate issuer and its content. A permissionless blockchain allows anyone to issue or receive certificates, ensuring signatures and verification mechanisms are available in perpetuity as long as at least one copy of the database is running. This allows anyone to verify the authenticity of a certificate, even if the issuer no longer exists.

New identification models are emerging, transforming from a single state-issued credential into knowledge-based aggregations of user-controlled attributes, such as reputation scores from social media, peer-to-peer sharing, and gig economy platforms (Hanson & Staples, 2017). Implementations of Blockchain Technology in Education Digital documents, often in proprietary formats, can be difficult to verify without the right software. Digital signatures come in various formats with various security levels, and not all of them are legally valid. Email sharing is not safe, so proprietary transmission infrastructures are needed for sensitive documents. Digital documents can also be spoofed by sophisticated users.

Blockchain technology can enhance the digital certification ecosystem by transforming private digital certificates into automatically verifiable information accessible to third parties. This eliminates the need for students to share sensitive metadata, which often includes private information. It also allows students to approach academic institutes and employers discreetly, with only the information marked as public during the proof-generation process. Software organizations can facilitate access to the Blockchain for students and badge issuers, ensuring data continuity and no lock-in with one solution. This can also foster social inclusion and entrepreneurship without requiring a centralized authority to verify the validity of Aglietti (2017).

2.3 Overview of the Blockchain Architecture at a Glance

Certification is a statement of facts with signatures as proof. Blockchains store cryptographic hashes or claims, ensuring the validity of the certificates. A blockchain is a system that connects sequential blocks of transactions, enabling the efficient and secure execution of transactions. To trade information across a private or public network, users need access to a blockchain wallet, a software application that mediates between users and the blockchain. Some wallets allow direct transactions without a third party, while others are managed by third parties. Users must install blockchain software on their devices in order to participate in consensus transactions. Without the need for a third party, anyone can transact directly within the network. A centralized authority determines who can run a node on permissioned blockchains. Blockchain transactions are tamper-proof, with blocks linked together through cryptography, making them impossible to be altered or deleted once recorded and time-stamped.

The blockchain records transaction details, parties, metadata, and a cryptographic hash. This unique signature is used to verify transactions, highlighting discrepancies if content is altered, ensuring integrity and security. The network guarantees that all parties involved in a transaction provide their consent before adding a new record, and only after verifying the capacity of the parties involved, the transaction is completed. The blockchain, running on thousands of computers, is mathematically verified to be identical to all others, making it impossible to 'hack' records by controlling over half of them. Public blockchains like Bitcoin and Ethereum make it nearly impossible to destroy the ledger entirely, as it would require deleting every copy worldwide.

2.4 Blockchain and Its Impact in the Education Sector

By incorporating core arguments from various studies (Morabito, 2017; Hanson & Staples, 2017; Gupta, 2017; Piscini et al., 2016), a set of principles can be proposed to understand the social value of blockchain technology in the field of education. Blockchain literature promotes *self-sovereignty*, enabling individuals to control their online identity, a concept that is particularly relevant in education, where learners can manage and share their credentials. (Lilic, 2015; Smolenski, 2016; Lewis, 2017).

The term "credential empowerment" is gaining prominence in the educational field, empowering learners to manage and share their credentials independently, without relying on institutions. Citizens gain self-authority over the online sharing of personal data and identities, allowing them to release information in exchange for desired services without relying on third-party validation. Smolenski adds, "The credential has emerged as a transnational, interdisciplinary signal of capability and skill in an environment where other characteristics – language, nationality, religious identity – cannot be presupposed" (Smolenski, 2017). Digital identity, a human right, is facing challenges due to private algorithms exploiting user data, altering our perception of identity and trust. Blockchain technology, based on cryptography, enables users to manage their personal data on their devices, thereby addressing identity gaps and facilitating efficient data storage and provision. Full digital self-sovereignty may shift from sharing permanent identities to verifying claims, requiring only relevant information for transactions. This system allows parties to verify individuals' age, education, and citizenship. Blockchain technology offers a potential solution to institutionalized trust issues by using decentralized public ledgers and cryptographic algorithms. These technologies automate the process of verification of the digital certification, without the involvement of third parties. Blockchain technology provides transparency, ease of sharing, and visibility in transactions with multiple organizations. It requires a public identifier for decoding and tracking and automatically authenticating transaction data. An immutable record is a document that cannot be changed after its creation. Blockchains, because of their immutability, provide security, secrecy, integrity, and availability. They are durable and irreversible, making transactional security and confidentiality virtually unassailable. This enhances trust in data integrity and decreases fraud opportunities. To be regarded as legitimate, all parties must agree on the transaction's legitimacy, which is widely distributed, ensuring that no single actor may be an arbiter of truth. Blockchain technology enables peer-to-peer consensus algorithms to record and verify transactions without a third party, potentially reducing costs, delays, and complexity. It can also guarantee authenticity across institutional boundaries, allowing parties to focus on new ways of authenticating records and transactions. Decentralization of the Internet would place more control on users' devices.

2.5 Certifications

Certifications give evidence of learning outcomes, teaching competence, learner involvement, meeting quality criteria, and accreditation body authority for educational organizations or courses. The development of new educational pathways is hampered by obsolete certification systems, particularly for those without access to high-level education. They frequently disregard informal and non- formal education, making it difficult to integrate learning into jobs and build beneficial after-school and after-work activities (Schmidt, 2017).

Components of a digital certification for educational qualification – A Certification is a statement from one party to another confirming the truth of a set of facts: (i) a claim asserts the truth of a set of facts, such as a learner acquiring a skill or a professor having sufficient knowledge; (ii) an issuer verifies this claim, and evidence supports it; (iii) a certificate attests the identity of the issuer, the recipient (learner or professor), the claim, and the evidence. It contains a unique signature, a symbol, a stamp, an image, or a code that only the issuer can affix, confirming their identity.

Processes Associated with Certification – There are three distinct processes: (1) *Issuing*: the process entails writing claim, issuer, evidence, receiver, and signature onto a certificate, which is usually stored in a centralized or produced by the user database; (2) *Verification*: a third-party verifies the authenticity of a certificate through three modalities: (a) using security features built into the certificate, (b) verifying with the original issuer, or (c) consulting a centralized database of claims. This process involves checking the authenticity of seals, special security papers, signatures, and centralized databases of claims. (3) *Sharing*: Certificate sharing entails transmitting the certificate to a third party, publishing it in a public registry or store it where the digital document is accessible to everyone.

Certificates are used in education to recognize various learning experiences such as (1) achieved total learning; (2) single learning units; (3) specific experiences, (4) skills acquisition; (5) excellence criteria; (6) competence levels and educational qualifications. Examples include university-leaving certificates, attendance certificates, mobility experience certificates, degree certificates, ECTS credits, work-experience certificates, apprenticeship certificates, and examination certificates or grade-cards (Grech et al. 2017). These certificates serve various purposes, such as recognizing the completion of specific learning experiences, attaining excellence criteria, and acknowledging competence levels. Certification in education is a motivational tool, gamifying learning by awarding certificates for achieving intermediate goals, which has been shown to improve concentration, recall, and overall learning outcomes (Gibson et al., 2015). The continuous formative assessment and certification program has been proven to enhance concentration, recall, and overall learning outcomes, as well.

2.6 *Blockcerts: An Open Standard for Blockchain Educational Certificates*

The MIT Media Lab and Learning Machine have created Blockcerts, an open standard for issuing and verifying credentials on the Bitcoin blockchain. The community aims to promote its adoption as the global standard for blockchain records. Blockcerts is a free, open-source standard for issuing and verifying digital records, allowing users to develop their own applications without any credits or royalties. It is used by organizations, start-ups and individuals worldwide. The platform's open-source nature aims to avoid standards wars and vendor lock-in, promoting interoperability and wide implementation. Blockcerts address the challenge of data silos, allowing for a more efficient and secure system. Students and work mobility benefits personal growth and organizations, but challenges still persist in the recognition of qualifications, including costs, time, and authenticity verification. Current systems require paper documentation, generating costs and bureaucratic issues. Blockchain technology may solve such issues: a certificated management and sharing system using decentralized standards and blockchain technology operates as follows: the graduates successfully verify their bachelor's degree with a foreign public institution using blockchain technology and PKI (public key) tools, ensuring authenticity, free of charge. Blockchain technology significantly enhances stakeholder efficiency, making it faster, safer and cheaper, thereby addressing the critical issues related to portability and study recognition and proving to be a powerful and effective technology.

2.7 *The Italian Blockchain Context*

Several European Union member states are implementing blockchain technology in vocational education and training, aiming to enhance citizens' mobility. In 2018, the European Blockchain Partnership was signed to optimize and harmonize blockchain technology in the European context. The 30 member states will focus on the strategic use cases, including qualification recognition, reinforcing the technology's strategic role in education and training policies. In April 2019, the University of Padua and the University of Milan-Bicocca, with Cineca's technical support, digitized their degrees, promoting the European Digital Credentials Initiative and Blockchain Observatory. Blockcerts certifications simplify credit uploads, removing administrative bottlenecks. This decentralization could improve availability, job market accessibility, and aid for migrant students. This universally feasible approach enables talent exchange, promotes human rights, and ensures degree verifiability. The innovative technologies offer benefits such as reduced duplicate certificate costs, instant verification of degrees, and improved job screening. The digitization of degrees makes them verifiable, flexible, and accessible to potential employers. Learning paths are based on networks, demonstrating knowledge through detailed syllabuses. The companies support the concept of digital credentials, allowing universities to pursue innovation and reduce administrative operations while ensuring the security, authenticity, and verifiability of degrees. Cineca, the largest Italian computing center, aims to integrate existing systems that have been digital repositories for didactics, students, and degrees.

3 Methodology

The focus of the research was placed on the study and observation of the practical application of blockchain technology in the higher education system, describing its current application, the benefits it can generate and the implications it might have in the future. The cases were categorized into four macro-categories of "solutions" of blockchains for technology application in degree recognition: (1) Roma Tre University; (2) DiploMe; (3) Blockchain4EDU – BCDiploma; (4) Bestr.

4 Results

This paper examines blockchain technology case studies for degree recognition and other certifications in education, documenting and generalizing application solutions. Criteria include the blockchain approach's representativeness, potential interest from practitioners, and sufficient data and information availability.

Table 1 Summary of the case study of blockchain adopted within a project at Roma Tre University

Description	Field	Approach to the Blockchain
The workers are empowered to own their professional identity through a platform for registration and sharing of their certificates	Professional and education degrees	Shared registry approach

Source: Authors' elaboration

4.1 Roma Tre University (Table 1)

Professional identity is often overlooked from the perspective of workers, lacking features like portability, ownership, and visibility of qualification titles. An accurate analysis of professional and training needs, despite the existence of digitized solutions aimed at representing the professional identity, is also difficult to perform. Roma Tre University is aiming to revolutionize the world of work by empowering workers with full access to and control over their professional data, digitizing it and allowing transparent sharing between public and private entities. Public and private entities can be employers or accredited agencies for employment services, as well as universities, and national institutions such as INPS (National Social Insurance Agency) or INAIL (National Institute for Insurance against Accidents at Work). The project aims to create the BE MAN platform (acronym of the technological functions of which it consists) to meet job offers and demand using a technological infrastructure consisting of BigChainDB, MongoDB, Mathematica, Angular, and Node, with a consensus mechanism based on the Tendermint and Byzantine Fault Tolerance mechanism. This case study involves multiple blockchain participants, including employment centers and third-party associations, aggregating data from various sources to create a shared register for efficient work-seeking and employment-offering interactions, allowing citizens to access services through employment centers. The anonymization of employees personal data is considered to be compliant with privacy requirements, but the implementation of the right methodology is still in progress.

4.2 DiploMe – Proponent (Table 2)

CIMEA – DiploME (https://www.cimea.it/EN/pagina-diplome) – MUR: CIMEA is the official Italian center within the *NARIC* (National Academic Recognition Information Centres) – network of the European Union and the *ENIC* (European National Information Centres) – network of the European Council and of UNESCO (Table 2).

DiploME's goal is to eliminate barriers to the portability of qualifications acquired through programs like Erasmus+ and International Exchange, promoting automatic acknowledgment and direct verifiability of diplomas and certifications

Table 2 Summary of the case study of blockchain DiploMe

Description	Field	Approach to the Blockchain
Citizen-oriented solution with over 15,000 active wallets, aiming to create a global ecosystem for managing qualifications and certifications	Higher education qualifications,academic degrees, VET qualifications (Vocational Education and Training), corporate certifications, micro-credential and certified skills	Shared registry – permissioned blockchain

Source: Authors' elaboration

across different systems. The solution aims to enhance the recognition process of academic and professional qualifications and certifications, thus reducing costs for users and public administration. The DiploME service was developed based on several fundamental elements: data quality and portability, safety, minimization, control and system opening. The DiploME system is tailored to specific user types: university students, post-graduate students, training course participants, and professionals with certifications. Academic institutions can utilize the DiploME ecosystem throughout the student's career, recording exams and curriculum courses on appropriate wallets. Stakeholders can issue non-academic qualifications, like certifications for specific computer training programs, and certification authorities can issue qualifications like ENIC-NARIC. DiploME is implemented using the Ethereum private version. Users of the DiploME service own public and private keys for data management. The private key provides full control over the wallet and the data. The holder of the keys also has *smart contracts* to manage qualifications and certifications. Certification authorities have public and private keys as well. The DiploME Wallet utilizes *smart contracts* to store certifications, which can be categorized into three privacy profiles. An authorized certification authority can write certificate information according to the user's profile requirements, depending on the desired level of privacy. The solution is currently active and more information can be found on the DiploME or CIMEA websites.

4.3 *Blockchain4EDU – BCDiploma – Proponent (Table 3)*

Blockchain Certified Data https://www.bcdiploma.com. The case study is analyzed using a standard structure to compare the obtained results (Table 3).

BCDiploma, a French project launched in 2018, aims to create a distributed platform for digital credentials and micro-credentials, combining blockchain technology with high encryption. It is a digital service designed to contrast the falsification of university diplomas and certifications. It uses Ethereum's blockchain technology to issue certificates, allowing graduates to prove their diploma's authenticity through a simple URL throughout their academic or professional lives. BCDiploma uses EvidenZ, an open-source framework for data implementation on-chain, aligned with privacy principles for data privacy. EvidenZ is designed for decentralized

Table 3 Summary of the case study of blockchain Blockchain4EDU – BCDiploma

Description	Field	Approach to the blockchain
It is a working group aiming to establish a shared blockchain platform for university degrees based on the BCDiploma case study for easy and safe certifications	Education degrees can be extended to professional, administrative, and personal certifications	Shared registry

Source: Authors' elaboration

implementation, allowing schools to operate autonomously through the consumption of a dedicated API. The BCDiploma project, with over forty clients worldwide, is still active and has helped establish numerous collaborations with schools and universities. According to the research carried out, BCDiploma is very widespread among institutions, bodies, and universities in French or Francophone countries. There are no registered Italian institutions that rely on BCDiploma, except for ESCP, which has more locations scattered around the world, including Italy: at the beginning of the 2021/2022 academic year, ESCP Business School has implemented an ambitious and innovative blockchain project. Once the cycle of studies comes to its end, the graduates of the prestigious business school will have a 100% digital and tamper-proof version of their precious diplomas. ESCP stated at the start of 2022 that all students entering its 27 specialized master's programs will receive a blockchain-certified diploma. Over 700 alumni now have access to a digital, forgery-proof copy of their diploma. By providing graduates with a unique URL link, which can also be presented as a QR code incorporated into a resume, ESCP Business School has digitized and streamlined the process of delivering certified attestations. Graduates will be able to confirm the authenticity of their diploma with a single click. The links are streamlined for integration into, for example, an online professional profile on social networks using a "mobile-first" approach (smartphone compatible).

Another interesting case study to mention is ETS Global, the TOEIC English test provider. With the advent of digitization in education and continuing professional education, the necessity to conveniently and securely communicate TOEIC® test scores has increased. As a matter of fact, the provider now offers applicants the option of receiving their score via a paperless score report. This technical advancement reflects an aim to improve the application process by converting the paper exam for the TOEIC® Listening and Reading test in the Public Program into a secure digital service. The service is the outcome of a large digital transformation program led by ETS Global's teams, which includes replacing paper flows with blockchain digital credentials and completely automated operations.

4.4 Bestr – Proponent (Table 4)

CINECA website: https://bestr.it/ (Table 4)

Table 4 Summary of the case study of blockchain Bestr

Description	Field	Approach to the Blockchain
Bestr can be used to represent a person's skills as digital credentials, and is dedicated to all University people, such as students, teachers, academic professors or technical- administrative staff	It includes hard and soft skills, certificates of participation, official educational certifications, and any digital credential that can be represented	Timestamping

Source: Authors' elaboration

Bestr is an Italian digital platform for certification and skill enhancement developed by Cineca, an Italian inter-university consortium and the largest Italian computing center, to record and enhance users' skills. The platform integrates Open Badge and Blockcerts. Blockcerts enables organizations to issue digital certificates signed by the issuing organization, allowing users to showcase their skills to companies, universities, and those involved in employee search or training, facilitating interaction between qualification holders and companies. Bestr uses the Blockcerts standard to create digital certificates with blockchain authenticity verification, extending Open Badge with notarization and enabling digital credentials to be issued in Open Badge or Open Badge plus Blockcerts. Bestr offers qualifications for various types of assignment. It is an innovative solution for companies, universities and other entities that allows them to publish and award recognitions to their users. Bestr badges let organizations display various competences, allowing users to search for profiles holding specific badges. These badges can be displayed in digital wallets, allowing companies to target specific competencies and qualifications. With Bestr, universities can expand the range of qualifications they issue, increasing student competitiveness and flexibility in their educational institutions. Through Blockcerts, an open standard, interoperability between solutions based on the same standard is guaranteed, enabling certifications to be used in several applications. The platform is integrated with university and project entities' information systems, enabling automatic digital credential issues for the education, research, and labor market sectors. Bestr uses *Ethereum* for notarization Blockcerts, a permissionless blockchain that can be issued and verified online by users without personal access to the Blockchain, agnostic in technology. Blockchain stores user qualification data as a digital *fingerprint*, not as personal information. This allows for the verification of the authenticity of initial data, as the hash of the original data within the blockchain is not traceable back to the original qualification information. The first project in production was that of the University of Milan – Bicocca, within which badges were issued representing the degrees of graduate students. In June 2019, 225 badges were assigned. In July 2020, Bestr is issuing Blockcerts for 56 courses of study at the University of Milan-Bicocca and 150 courses at the University of Padua.

4.5 University of Turin (UniTo)

In the academic year 2022/2023, the University of Turin implemented Blockcerts in its IT Systems, Website Management and E-learning Division with a scheduled release of Open Badges for a starting group of 380 second-cycle graduates in *Business Administration, Economics and Management, Economics of the Environment, Culture, and Territory,* and *Statistical and Economic Methods for Decisions* by the end of December 2023. The University of Turin uses the platform, for digital credentials, Bestr (https://bestr.it/?ln=en), one of the five fully certified products in the world for issuing and managing digital credentials. The Blockcerts standard builds on the Open Badge 2.0 standard and connects it to blockchain technologies. Bestr writes its blockcerts on the public Ethereum blockchain, using hashing and Merkle-tree systems to ensure the lowest number of transactions while maintaining the maximum level of anonymity. By 2022, over 120 organizations and 46 European universities rely on this platform. Furthermore, Bestr obtained the displayer and host certifications of 1EDTECH – an international consortium in digital education (https://www.1edtech.org/).

In the frame of the project "Ecosystem for (micro) digital credential UniTo: open badge, blockcerts and blockchain", in 2023, the University of Turin has faced this innovative and digital transformation with blockcerts notarized in its own blockchain. The IT Systems, Web Management and E-learning Division had to deal with several steps before refining the blockcerts. First of all, it was necessary to have several meetings to promote technological innovation with a bottom-up approach, starting with the administrative technical staff and then involving the teaching managers and the institutional positions, such as the head of the master's courses, with the help of the Deputy Directors for Teaching of the departments involved. The promotion force was born within the School of Management and Economics, with the support of the two Deputy Directors for Teaching of the departments of Management "Valter Cantino" and of Economics, Social Studies Applied Mathematics and Statistics (ESOMAS)," plus the staff of the E-Learning Division. The same promotion campaign for blockchain culture was conducted symmetrically and with the same bottom-up approach at the School of Law Politics and Social-Economic Sciences with the collaboration of the department of Economics and Statistics "Cognetti de Martiis". The IT Systems, Web Management, and E-Learning Division chose for all the Blockcerts a graphic design that reflects the visual identity of the University of Turin.

In the upper part of the Blockerts is shown the corresponding open badge for the Master's Degree (Fig. 1). There are the name and surname of the graduate, the date of the achievement of the academic degree, a brief description of the course program, and an indication of the level of the EQF (European Qualification Framework). In this case, the seventh level is indicated for the Master's Degree and the eighth level for PhDs. In addition, there is the digital signature of the rector. The indication of the level recognized in the EQF allows the learner beneficiary to make his degree immediately recognizable and comparable to all European states and stakeholders.

Fig. 1 Visual of open badge for Master's Degrees UniTo. (Source: Authors' elaboration)

Since 8th May 2023, the University of Turin is in the Italian register of Blockcerts issuers (https://bestr.it/issuerspublickeyregistry/view) and public identifiers with its own pubkey to issue the blockcerts notarized in the blockchain of the university. Firstly, Esse3 – the platform for managing the students' careers, used and managed by the Student Services Offices – communicates the educational qualification acquired by the graduates, upon the standard experience API. Secondly, Bestr recognizes the education degree through the assignment of the Open Badge for the Master's Degree. Bestr invites the graduates to claim their blockcerts. The graduates communicate their public address directly on Bestr or via the mobile Wallet App. Bestr creates the blockcerts and it notarizes the blockcerts on the blockchain: the transaction takes place from the public address of the university. Once the notarial certification has taken place, the graduates receive a notification. To get it, they need to check into their university e-mail box. They will receive an email from noreply@bestr.it with an invitation to access the Bestr platform and to collect and activate their own Blockcerts. They need to download the Blockcerts Wallet App (available on Google Play or the App Store) on their smartphone or mobile device. The graduates will have a single recognition number, which can be entered in their own Blockcerts. The App is used to manage their own ID for the Blockcerts and to give it to the organizations they want to authorize to appoint Blockcerts. Moreover, they can also use it to store and share all their Blockcerts. Once the graduates have their ID, they have to communicate it to the organization they want to provide the Blockcerts with. To do so, they have to add the organization to the authorized issuing institutions using the App: by scanning the QR code shown on the Bestr page in English with the smartphone on which the app is installed, the app will open and the issuing institution will be added. Otherwise, they have to choose to add the issuing institution manually from the app settings and copy the two codes provided by the page: the recognition URL and the one-time code. After this procedure is complete, Bestr UniTo will have all the necessary information to write the Blockcerts and save them in the blockchain. In 2024, the University of Turin is willing to extend the project to Veterinary Medicine and Agricultural Sciences graduate students. The

Bestr project is making significant progress, with universities already employing Issuers and for creating an ecosystem through an open, interoperable standard like Blockcerts.

The blockcerts notarized in the blockchain are very useful in the field of student mobility, recruiting, and screening e-CVs for public competitions, for jobs in higher institutions and research centers in the absence of the institution that issued the certificate.

In order to check the validity of the blockcerts, there are some open-source universal validators that verify the integrity of the digital document (https://www.blockcerts.org/) and download the record in .json. Blockcerts use open standards for recipient-centric certificates. The Blockcerts community supports and works to improve the following standards: Open badges from IMS, Verifiable W3C Claims, Linked Data Signatures from the W3C, Rebooting the Web of Trust Decentralized Identifiers (W3C). The open-source components of Blockcerts adhere to the aforementioned criteria, illustrating how such standards can be implemented in a functional system. This includes developing answers to problems that arise in a real-world system, such as the revocation of participant credentials in a decentralized, self-sovereign mode, and other lifecycle issues, like key management. The blockchain is an immutable and distributed transaction store.

5 Conclusion and Future Developments

After the issue of blockcerts in the blockchain to 380 graduates of the four Master's Degree courses of the University of Turin and the corresponding equalization with the graduate sessions, the university has programmed the application of the blockchain for the Master's Degrees in Agricultural Sciences and Veterinary Medicine and for some PhDs in a scalable manner. The task force employed in the implementation, dissemination, and use within the ecosystem of the university's digital credentials is composed not only of IT technicians but also of different professional profiles, from teaching experts to people working on institutional communication and management of social media. Only by combining the skills and professionalism present on campus it will be possible to implement and use blockchain in the correct and profitable way for bachelor's, master's, and other academic qualifications. Blockcerts certifications could improve international mobility by decentralizing the loading of educational credits and micro credentials in the students' track record. Intercommunication between central offices of the different institutions is a major obstacle to the free international movement of students. The verifiability of titles is crucial for forced mobility and protecting human rights. Some Italian universities, such as Milano Bicocca and Padua, have faced cases of migrant or refugee students declaring qualifications no longer available or totally inaccessible, and embassies and specialized agencies struggle to help. By storing self-verified certificates in the cloud, these problems could be permanently solved. The University of Turin will use this technology to welcome international students and promote mobility for study, training and research.

The study surely has some limitations since it is an exploratory research area in its early stages. This chapter employs an exploratory, qualitative approach, utilizing a limited number of case studies. However, the study illustrates how digital certifications could revolutionize the current dynamics of the job market and reduce fraud, and how self-verified certificate storage in the cloud could solve some of the criticalities that the institutions, academic and nonacademic, are facing in a continuously evolving world.

The future of this innovation depends on the widespread acceptance of digital certifications by other universities and institutions. These certifications can also be used for global talent exchange and human rights protection, so hopefully they might be a useful instrument to improve national and international cooperation.

References

Aglietti, A. (2017). *Proof-of-Knowledge: Same Blockchain, different story*. Available at: https://medium.com/growbit/proof-of-knowledge-efc138f2a17c

Calandra, D., Secinaro, S., Massaro, M., Dal Mas, F., & Bagnoli, C. (2022). *The link between sustainable business models and Blockchain: A multiple case study approach*. https://doi.org/10.1002/bse.3195

Gibson, D., Ostashewski, N., Flintoff, K. et al. (2015). Digital badges in education. Educ Inf Technol 20, 403–410. https://doi.org/10.1007/s10639-013-9291-7

Grech, A., & Camilleri, A. F. (2017). *Blockchain in education*. Inamorato Dos Santos, A. (Ed.). EUR 28778 EN; https://doi.org/10.2760/60649

Hanson, R. T., & Staples, M. (2017). *Distributed ledgers, scenarios for the Australian economy over the coming decades*. Canberra, Commonwealth Scientific and Industrial Research Organization.

Lewis, R., McPartland, J. S., & Ranjan, R. (2017). Blockchain and financial market innovation. *Economic Perspectives, 41*, 2–12.

Lilic, J. (2015). *uPort; A glimpse into a next generation self sovereign identity system*. Available at: https://www.linkedin.com/pulse/uport-glimpse-next-generation-self-sovereign-identity-john-lilic

Morabito, V. (2017). *Business innovation through Blockchain. The B3 perspective*. Springer. https://doi.org/10.1007/978-3-319-48478-5

Piscini, E., Guastella, J., Rozman, A., & Nassim, T. (2016). *Blockchain: Democratized trust. Distributed ledgers and the future of value*. Deloitte University Press. Available at: https://www2.deloitte.com/content/dam/Deloitte/uk/Documents/technology/deloitte-uk-tech-trends-2016-blockchain.pdf

Raimundo, R., & Rosário, A. (2021). Blockchain system in the higher education. *European Journal of Investigation in Health, Psychology and Education, 11*(1), 276–293. https://doi.org/10.3390/ejihpe11010021

Reis-Marques, C., Figueiredo, R., & de Castro, N. M. (2021). Applications of Blockchain technology to higher education arena: A bibliometric analysis. *European Journal of Investigation in Health, Psychology and Education., 11*(4), 1406–1421. https://doi.org/10.3390/ejihpe11040101

Schmidt, J. P. (2017). *Credentials, reputation, and the Blockchain*. Available at: http://er.educause.edu/articles/2017/4/credentials-reputation-and-the-Blockchain

Smolenski, N. (2016). *Identity and digital self-sovereignty. A new paradigm for sovereignty on the high Seas*. Available at: https://medium.com/learning-machineblog/identity-and-digital-self-sovereignty-1f3faab7d9e3

Sousa, M. J., Dal Mas, F., Gonçalves, S. P., & Calandra, D. (2022). AI and Blockchain as new triggers in the education arena. *European Journal of Investigation in Health, Psychology and Education, 12*, 445–447. https://doi.org/10.3390/ejihpe12040032

Online Resources Websites

Gupta, V. (2017). *European Parliament Blockchain presentation*. Available at: https://www.youtube.com/watch?v=fCYT9KWoldI

http://www.ponricerca.gov.it/media/396162/rapporto-di-disseminazione-blockchain.pdf

https://blockchain.mit.edu/

https://blog.bestr.it/en/2018/10/15/blockchain-and-open-badges-bestr-becomes-digital-credentials-ecosystem

https://blog.bestr.it/en/2019/06/13/blockcerts-recruiters-and-evaluators-guide

https://blog.bestr.it/en/2019/06/13/reading-verifying-and-sharing-blockcerts-bestr

https://blog.bestr.it/en/2019/09/23/digitizing-university-degree-how-blockchain-going-redefine-bachelors

https://community.blockcerts.org/

https://escp.eu/news/MS-MSc-diplomas-certified-with-blockchain-technology

https://ethereum.org/en/

https://wiki.p2pfoundation.net/Blockchain

https://www.bcdiploma.com/en/blog/itw-laurence-carlinet-executive-director-france-new-partnerships-ets-global-2021-11-09

https://www.blockcerts.org/

https://www.digitalcredentials.unito.it/english

https://www.media.mit.edu/projects/media-lab-digital-certificates/overview/

Inamorato dos Santos, A. (2017). *Blockchain in education – European Commission's JRC report preview, Blockchain in Education Conference, Groningen*. Available at http://bit.ly/2lsyvDb

Smolenski, N. (2017). *Blockchain records for Refugees*. Available at: https://medium.com/learning-machine-blog/blockchain-records-for-refugees-bd27ad6e6da1

Competence and Soft Skills in Higher Education – Innovation in Evaluating the Learning Outcomes of University Students: The TECO Project

Francesca Di Virgilio

1 Introduction

In this historic period, characterized by worldwide recession phenomena due to pandemics and war, among the traditional objectives proposed in the higher education's instructional activities, the additional aim of educational authorities is to promote safe and effective learning environments for all students, as well as improving teaching skills (ONU, 2021).

Today, higher education institutions are being evaluated by the manners in which they react to the social and economic needs of society, that is, how they promote the modernization through the enhancement of competences and soft skills, within their education systems, and the implementation of a collaboration model with the entrepreneurial system, in order to align the academic world to the challenges of the labor market (Di Virgilio & Asunka, 2019; Soliman et al., 2023). In a world characterized by constant innovation and rapid change, human resources represent a driver to guarantee organizations a defensible and lasting competitive advantage, and the intellectual capital brought to organizations is becoming increasingly important (Mondal et al., 2022). The intellectual capital, formed by the knowledge, skills, and competences of the young generation, is defined as a critical success factor as it is strictly determining the company's performance (Torre et al., 2020). Companies that adopt this philosophy in HR management attach great importance to the enhancement of these resources and therefore devote a lot of effort to managing and integrating them with the corporate culture and strategy (Mondal et al., 2022). In companies that operate in the public and private contests, that adopt a constitutive rather than an instrumental approach, people take an active role in creating a

F. Di Virgilio (✉)
University of Molise, Campobasso, Italy
e-mail: fradivi@unimol.it

A. de Bem Machado et al. (eds.), *Digital Transformation in Higher Education Institutions*, EAI/Springer Innovations in Communication and Computing,
https://doi.org/10.1007/978-3-031-52296-3_12

competitive advantage, since they are considered bearers of an autonomous value and capable of developing skills and relationships (Wright et al., 1994).

Employability skills refer to the skills that are required, sustaining, and doing well on a job. These skills and characteristics allow graduates to resolve problems, develop deference, and eventually turn out to be robust representatives designed for the business (Tang, 2019; Yorke, 2006).

However, the main aim of higher education's instructional activities outspreads away from guaranteeing that graduates are employable. The courses covered by higher education institutes seem to be more theoretical.

In this scenario, the higher education system aims at building the capacities of students; it shall aim at effectively developing competences and soft skills among the students to prepare them for the job market. They are useful in making students remember and understand the theories but fail to apply and analyses those theories in practice.

Understanding teacher factors that might drive student academic performance is becoming an important topic of research in educational researches. In general terms, both researchers and educational authorities are interested in identifying elements that could contribute to improve the quality of education (Varouchas et al., 2018) and determine the quality of the education and learning process (Soliman et al., 2021). As such, HEIs are required to provide a suitable educational climate and environment through which to ensure that cultural diversity is supported, the identified core values of the institution are well promoted, and adequate career development programs for students are well provided. (e.g., Daniel, 2010; Giacalone & Jurkiewicz, 2003; Jurkiewicz et al., 2004). Teachers drive innovation in pedagogical methodologies and encourage students' involvement (Khoo-Lattimore, 2018a, b; Walder, 2014).

The use of data provided by academic and learning analytics will allow the implementation of effective strategies to improve the quality of education. In this sense, analytics are a powerful mechanism to help students, instructors, teachers, designers, and developers of learning systems to better understand educational processes and predict students' needs and performance (Klašnja-Milicevic & Ivanovic, 2018; Zilvinskis & Willis III, 2019), and to ultimately achieve exceptional learning outcomes (Shephard, 2008).

2 The Case Study: TECO Project

In the European scenario, Italy on higher education held by the Ministry of Higher Education, Research, Science and Technology aims to contribute to the improvement of the quality of the training process by activating internal self-assessment mechanisms in the academic world. In 2012, ANVUR[1] launched: the TECO project

[1] Italian national agency for the evaluation of universities and research institutes.

(TEst on Competences). The TECO project, through the construction of indicators: Literacy,[2] Numeracy,[3] Problem Solving[4] and Civics,[5] estimates the levels of skills of university students. TECO is therefore consistent with the most recent European guidelines,[6] relating to the European Higher Education Area (EHEA), which promote student-centered teaching, accompanied by the analysis of the learning outcomes.

TECO is an integral part of the Self-Assessment, Periodic Assessment, Accreditation (AVA) system in the higher education institutions. The quality assessment activities of the outcomes of the training offer are also regulated by the constitutive Presidential Decree of the Agency. In particular, the efficiency and effectiveness of teaching activities are assessed on the basis of international quality standards, also with reference to the learning outcomes of students and their adequate integration into the world of work.

The TECO project provides for the definition and administration of tests capable of detecting the transversal (TECO-T) and disciplinary (TECO-D) skills acquired by students during the 3-year course of study. The TECO-T tests were carried out by the Agency with the collaboration of groups of experts consisting mainly of university professors, following a top-down process. The detection of disciplinary skills, on the other hand, makes use of the commitment of disciplinary groups assisted by ANVUR: after the analysis of the Course Sheets, each disciplinary group identifies the core disciplinary contents and declines them with respect to the 5 Dublin Descriptors. TECO's is the first experience of its kind in Italy in the field of higher education and is rapidly gaining the interest of evaluators and scholars from other countries.

TECO surveys include tests that are administered to all students who agree to participate. Questions and answers, mostly closed-ended, are predetermined, thus eliminating the arbitrariness and variability of the evaluator.

TECO tests are administered to the computer, using a platform set up and managed remotely by CINECA, within classrooms identified by the universities. The time windows (September–December) are set by ANVUR in collaboration with CINECA[7]. The TECO-2020/2021 Survey has been carried out for the first time, on the transversal skills identified for all students were those of Literacy Numeracy, Problem Solving, and Civics organized in a single test lasting a total of 50 min. Two

[2] Reading comprehension.

[3] Understanding and solving quantitative logic problems.

[4] Understanding and solving simple problems.

[5] Civic knowledge – (citizenship rights, end of life rights, religious rights, etc.).

[6] European Standards and Guidelines for Quality Assurance, 2015.

[7] Cineca, which is a Consortium by 112 Italian universities and public institutions composed, is led by a Board of Directors composed of the rectors/delegates of the universities and public bodies members of the Consortium, and one delegate of the Ministry of Education, and the Ministry of University and Research. The Board of Directors is represented by the Chairman, while the CEO is responsible for the development, organization and management of the Consortium's activities.

parallel sets of TECO-T were provided, randomly assigned to students, and also within the same virtual classroom.

The results of the TECO test found that transversal skills draw on a generalist training background and can be also developed in the university course of study. Furthermore, they are comparable between universities and/or between courses.

TECO test offers study course information that allows self-assessment in order to:

- initiate or deepen the reflection on transversal skills in degree programs, on the basis of a broader information framework on the profile of their students;
- help to inform the analysis of the strengths of the training offered with regard to disciplinary skills and the reflection on possible interventions for improvement with respect to the design and implementation of the Degree Programs.

Moreover, previous studies showed that a degree of mismatch occurred due to the soft skills acquired by the graduates at higher education institutions are insufficient to use in their employment (Tang, 2019), the students' participation in TECO initiates a self-assessment process to improve their first entry into the labor market evaluating their levels of soft skills.

3 Conclusion

In spite of this growing importance, the set of competences is struggling to find defined areas, stable training models, classification, and accreditation systems that are capable of permanently incorporating soft skills into the curricular training paths of universities. TECO is designed for transversal skills learning, education, and the practical application of soft skills to enable HEIs involved to put their knowledge and resources at the service of their local/national/regional communities.

The final outcome expected from the project is promoting reform in HEIs to become catalysts of economic and social development in the European participating countries. Considering this scenario, and particularly in nascent ecosystems, entrepreneurial universities play a fundamental role by acting as catalysts and changemakers, since they can offer unique opportunities of connection between the research and the commercial worlds. By creating their own innovation hubs, in fact, they can have both a direct and indirect impact on the country actors, and can reach goals in terms of infrastructure upgrading, talent development, talent attraction, innovation development, rates of new firm creation, rates of startup investments, collaborations among actors, and development of a culture of entrepreneurship and innovation.

It is relevant particularly in Italy, despite the fact that the educational systems of the Region recognize the importance of the development of soft skills, both in the guiding documents and in the learning models, there is no common framework to address and make permanent the learning model and training standards. In other words, the development of learning of soft skills is applied as a generic objective of integral formation of the student, without sufficient explanation.

What is more, transversal competences have a cross-cutting character, regarding mostly the jobs characteristic of the Fourth Industrial Revolution. The EU points pout the relevance of this kind of skills in the Digital Education Action Plan (2021–2027), which aims to support the sustainable and effective adaptation of the education and training systems of EU member states to the digital age. The action is expected to contribute to Sustainable growth and jobs (build skills for life and work) supporting sustainable recovery in European countries.

In line with the strategic priorities of the Digital Education Action Plan (2021–2027). These goals will be fulfilled by implementing comprehensive interventions such as introducing innovative elements in the existing curricula of Italy HEIs and transferring innovative learning and teaching methods.

In this scenario, the TECO Project results in line with the soft skills priorities as it takes action on two key dimensions:

(a) The model of classification, learning, learning assessment, and recognition of soft skills. A model built around a library is extensible in any context and environment and does not require long paths of institutionalization and, instead, constitutes a transferable practice and can be improved continuously. In this respect, the project points at enhancing the teaching, assessment mechanisms for HEI staff and students, quality assurance, management, governance, inclusion, innovation, and knowledge base.
(b) The training of teachers and administrative staff, not only through a specific training program in each university but also through the construction of a model for the training of teaching staff on soft skills in each university. In this sense, the project reaches the scope of improving the training of teachers and continuous professional development in order to impact the long-term quality of the education system.

The TECO works as a priority: Sustainable growth and jobs (build skills for life and work) – *it* can support sustainable recovery in European countries, address traps, and increase competitiveness. The project is designed also to improve students' desire to become entrepreneurs and/or innovators; animate their entrepreneurial spirit; facilitate the creation of an entrepreneurial ecosystem; foster collaboration among students from all disciplines and faculty members; help students develop those soft skills that are increasingly being requested by companies that operate in innovative sectors; and promote Technology Transfer and Open Innovation.

References

Daniel, J. L. (2010). The effect of workplace spirituality on team effectiveness. *Journal of Management Development, 29*(5), 442–456.

Di Virgilio, F., & Asunka, S. (2019). Social Media for knowledge sharing in higher education institution in the sub-Saharan African context. In A. Stachowicz-Stanusch & L. Mercurio (Eds.),

The social issue in contemporary society: Relations between companies, public administrations and people. IAP – Information Age Publishing. ISBN 9781641135597.

Giacalone, R. A., & Jurkiewicz, C. L. (2003). Toward a science of workplace spirituality. In R. A. Giacalone & C. L. Jurkiewicz (Eds.), *The handbook of workplace spirituality and organizational performance*. M.E. Sharpe, Armonk.

Jurkiewicz, C. L., Robert, A., & Giacalone, R. A. (2004). A values framework for measuring the impact of workplace spirituality on organizational performance. *Journal of Business Ethics, 49*, 129–142.

Khoo-Lattimore, C. (2018a). The changing landscape of academic tourism research towards a hopeful academy. *Tourism Management Perspectives, 25*, 131–132.

Khoo-Lattimore, C. (2018b). The ethics of excellence in tourism research: A reflexive analysis and implications for early career researchers. *Tourism Analysis, 23*(2), 239–248.

Klašnja-Milicevic, A., & Ivanovic, M. (2018). Learning analytics—New flavor and benefits for educational environments. *Information Education, 17*, 285–300.

Mondal, S. R., Di Virgilio, F., & Das, S. (Eds.). (2022). *HR analytics and digital HR practices*. Palgrave Macmillan.

ONU. (2021). *Objetivos de Desarrollo Sostenible*. Available online: https://www.un.org/sustainabledevelopment/es/objetivos-dedesarrollo-sostenible/. Accessed 26 July 2021.

Shephard, K. (2008). Higher education for sustainability: Seeking affective learning outcomes. *International Journal of Sustainability in Higher Education, 9*(1), 87–98.

Soliman, M., Di Virgilio, F., Figueiredo, R., & José Sousa, M. (2021). The impact of workplace spirituality on lecturers' attitudes in tourism and hospitality higher education institutions. *Tourism Management Perspectives, 38*, 100826. https://doi.org/10.1016/j.tmp.2021.100826

Soliman, M., Sinha, R., Di Virgilio, F., Sousa, M. J., & Figueiredo, R. (2023). Emotional intelligence outcomes in higher education institutions: Empirical evidence from a Western context. *Psychological Reports*. https://doi.org/10.1177/00332941231197165

Tang, K. N. (2019). Beyond employability: Embedding soft skills in higher education. *The Turkish Online Journal of Educational Technology, 18*(2), 1–9.

Torre, C., Tommasetti, A., & Maione, G. (2020). Technology usage, intellectual capital, firm performance and employee satisfaction: The accountants' idea. *The TQM Journal, 33*, 1754–2731.

Varouchas, E., Sicilia, M. A., & Sanchez-Alonso, S. (2018). Academics' perceptions on quality in higher education shaping key performance indicators. *Sustainability, 10*, 4752.

Walder, A. M. (2014). The concept of pedagogical innovation in higher education. *Education Journal, 3*(3), 195–202.

Wright, P. M., McMahan, G. C., & McWilliams, A. (1994). Human resources and sustained competitive advantage: A resource-based perspective. *International Journal of Human Resource Management, 5*(2), 301–326.

Yorke, M. (2006). Employability in higher education: What it is – What it is not. In *Learning & Employability, series one*. The Higher Education Academy.

Zilvinskis, J., & Willis, J. E., III. (2019). Learning analytics in higher education: A reflection. *InSight: A Journal of Scholarly Teaching, 14*, 43–54.

Correction to: Use of ChatGPT in Higher Education: A Study with Graduate Students

Rogério Costa, Adriana Lage Costa, and Ana Amélia Carvalho

Correction to:
Chapter 7 in: A. de Bem Machado et al. (eds.),
***Digital Transformation in Higher Education Institutions*,**
EAI/Springer Innovations in Communication and Computing,
https://doi.org/10.1007/978-3-031-52296-3_7

The chapter was inadvertently published with wrong affiliation of the chapter author. This has been corrected and the affiliation matches the one that was originally submitted.

The updated version of this chapter can be found at
https://doi.org/10.1007/978-3-031-52296-3_7

A. de Bem Machado et al. (eds.), *Digital Transformation in Higher Education Institutions*, EAI/Springer Innovations in Communication and Computing,
https://doi.org/10.1007/978-3-031-52296-3_13

Index

A. de Bem Machado et al. (eds.), *Digital Transformation in Higher Education Institutions*, EAI/Springer Innovations in Communication and Computing,
https://doi.org/10.1007/978-3-031-52296-3

MIX
Papier aus verantwortungsvollen Quellen
Paper from responsible sources
FSC® C105338

If you have any concerns about our products,
you can contact us on
ProductSafety@springernature.com

In case Publisher is established outside the EU,
the EU authorized representative is:
Springer Nature Customer Service Center GmbH
Europaplatz 3, 69115 Heidelberg, Germany

Printed by Libri Plureos GmbH
in Hamburg, Germany